Quality Assurance Of Chemical Measurements

John K. Taylor, Ph.D.
Center for Analytical Chemistry
National Bureau of Standards
Gaithersburg, MD 20899

TABLE OF CONTENTS

I INTRODUCTION
II MEASUREMENT AS A PROCESS - CHEMICAL ANALYSIS AS A SYSTEM
III STATISTICAL CONCEPTS
IV MODELING
V PLANNING AND SAMPLING
VI METHODOLOGY
VII CALIBRATION
VIII QUALITY ASSURANCE - GENERAL ASPECTS
IX QUALITY CONTROL
X CONTROL CHARTS
XI QUALITY ASSESSMENT
XII CORRECTION OF ERRORS AND/OR IMPROVING PRECISION AND ACCURACY
XIII MEASUREMENT CAPABILITY - THE NATIONAL MEASUREMENT SYSTEM - TRACEABILITY
XIV REFERENCE MATERIALS
XV REPORTING DATA
XVI VALIDATION
XVII LABORATORY CERTIFICATION/EVALUATION
XVIII PLANNING QUALITY ASSURANCE PROGRAMS
XIX APPENDICES

   A (Reprint) - Quality Assurance of Chemical Measurements
   B (Reprint) - Sampling for Chemical Analysis
   C Guidelines for Evaluating the Blank Correction
   D Validation of Analytical Methods
   E Selected references

August 1984

I

INTRODUCTION

MEASUREMENT

"If you can measure what you speak of and can express it by a number, you know something about your subject; but if you cannot measure it, your knowledge is meagre and unsatisfactory . . . '

                                        Lord Kelvin

"The ability to measure is one of man's great capabilites"

                                        F. D. Rossini

"The trouble with measurement is its seeming simplicity"

                                        Anon

"Measure what is measurable and render measurable that which is not yet measurable"

                                        Galileo Galilei

                                        1564-1642

MEASUREMENTS NEEDED FOR

Basic Science

Identification of Problems

Solving Problems

Control of Production

Evaluation of Products and Services

       in

Virtually Every Area of Human Life and Welfare

               MEASUREMENTS ALLOW HUMANS

To:

    Describe
    Predict
    Communicate
    Decide
    Control
    React

               WHEN USED FOR DECISIONS

The Questions

    How Good?
    How Sure?

Are rightfully asked and must be properly answered

A GOOD METROLOGIST MUST BE ABLE TO:

    Make Good Measurements
    Define His/Her Measurements
    Defend the Measurements

        Technically Sound
        Legally Defensible

Interpret His/Her Measurements

                    Supportable

                    Value Judgments

                    Need to be Made

                    Quality

TWO KINDS OF QUALITY

Design Quality

    Design developed to meet a need

Conformance Quality

    The actual production of a product that meets the need

Quality is the Absence of Defects

    Deficient - General Inadequacy
    Defective - Wanting in some essential property, faulty

QUALITY PARAMETERS

QUALITATIVE IDENTIFICATION

"Certainty"

QUANTITATIVE ACCURACY

Probability

Confidence Limits

Positive identification is the first requirement for reporting data. Limits of uncertainty are needed to judge the confidence associated with the numerical result.

A LABORATORY IS EXPECTED TO BE ABLE TO SPECIFY THE QUALITY OF ITS DATA IN QUANTITATIVE TERMS. THIS REQUIRES THE EXISTENCE OF SOME DEGREE OF QUALITY ASSURANCE.

DEFINITIONS

Measure - to ascertain the extent, degree, quantity, dimensions, or capability with respect to a standard, hence to estimate.

Measurement

    A Process
    An Operation
    A Piece of Data

1. Act or result of measuring something
2. The extent, size, capacity, amount or quantity ascertained by measuring
3. A system of measures
4. Math - the correlation with numbers of entities that are other than members of aggregates

CHEMICAL MEASUREMENTS

Measurements of chemical substances

Measurements which chemists make to obtain information on chemical substances and chemical systems

The data obtained as a result of measurements on chemical substances or chemical systems

QUALITY ASSURANCE OF CHEMICAL MEASUREMENTS

OBJECTIVES

    To Assess Limits of Error in Measurements
    To Reduce Analytical Errors to Acceptable Levels
    To Reduce Amount of Work Needed to Obtain Reliable Data
    To Provide Basis for Intercomparison of Data

TWO CONCEPTS

    Quality Control
    Quality Assessment

BASIC REQUIREMENTS

    Understand the Nature of Errors
    Understand the Measurement System Used
    Develop Techniques and plans to Minimize Error

DEFINITIONS

QUALITY CONTROL - The overall system of activites whose purpose is to provide a quality of product or service that meets the needs of users; also, the use of such a system. The aim of quality control is to provide quality that is satisfactory, adequate, dependable, and economic. The overall system involves integrating the quality aspects of several related steps including: (a) the proper specification of what is wanted; (b) production to meet the full intent of the specification; (c) inspection to determine whether the resulting product or service is in accord with the specifications; and (d) review of usage to provide for revision of specifications.

QUALITY CONTROL - A system of inspections, testing, and remedial actions applied to a process or operation so that, by inspecting a small portion (a sample) of the product currently produced, an estimate can be made of its quality and whether or not, or what if any, changes need to be made to achieve or maintain a predetermined or required level of quality.

QUALITY ASSESSMENT - A system of activities whose purpose is to provide assurance that the overall quality control job is in fact being done effectively. The system involves a continuing evaluation of the adequacy and effectiveness of the overall quality control program (see quality control) with a view to having corrective measures initiated where necessary. For a specific product or service, this involves vertifications, audits, and the evaluation of the quality factors that affect the specification, production, inspection, and use of the product or service.

QUALITY ASSURANCE

QUALITY CONTROL/QUALITY ASSESSMENT

QUALITY CONTROL - Those procedures and activities developed and _implemented_ to produce a product/measurement of required quality.

QUALITY ASSESSMENT - Those procedures and activities _utilized_ to verify that the quality control system is operating within acceptable limits.

QUALITY ASSURANCE

TECHNICALLY SOUND AND LEGALLY DEFENSIBLE

There is a difference and both requirements need to be met. The first is necessary but not sufficient. It includes all those things that a careful competent analytical chemist would do. The second includes the proof that sound work was done. Quality assurance practices are central to both. They stress documentation which takes the burden off memory and can remove any suspicion or shadow of doubt of details of what was done. Such measures cannot give credence to poor technical work. In fact they can help to identify poor quality when it exists. But lack of documentation can make defense deficient if not impossible. While legal defense is often the issue, technical reliablity is always at stake. Quality assurance practices can promote this, as well.

QUALITY ASSURANCE is the name given to procedures by which one ASCERTAINS that INDIVIDUAL MEASUREMENTS are GOOD ENOUGH for their INTENDED PURPOSE.

The SHADOW OF DOUBT should be SUITABLY SMALL to permit

VALID SCIENTIFIC INFERENCE

EFFECTIVE PROCESS CONTROL

QUALITY PRODUCTS

INTELLIGENT ACTIONS

QUALITY ASSURANCE vs QUALITY CONTROL

Quality Control - Old Concept

    Widely endorsed
    All good workmen strive for quality outputs
    Quality recognized as necessary

Quality Assurance - Recent Concept

    Statistical control of quality

Resistance to QA

    Based on

- Improve technical methods so that no important quality variations remain

- Statistics have no proper place among scientific production methods

- If product is good, no inspection is needed; if not, inspection will not help

QUALITY ASSURANCE

OF A PRODUCT

- o For Conformance with Specifications
- o Involves

    - o Sampling a Defined Population
    - o Measurement of a Distinctive Property

        - o Variance of Measurement System often Considered Negligible

OF A MEASUREMENT

- o Data Output may be Considered as the Product
- o Typical Classes Include

    - o Calibrations
    - o Chemical Analysis

- o Involves

    - o Establishing statistical control
    - o Evaluating precision by repetition
    - o Evaluating bias by reference materials

PRODUCTION PROCESS QUALITY ASSURANCE

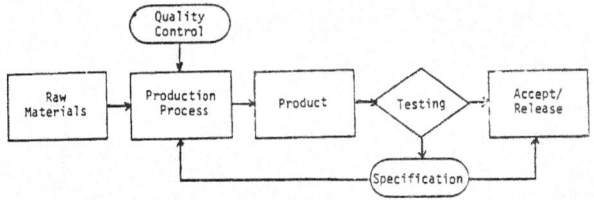

MEASUREMENT PROCESS QUALITY ASSURANCE

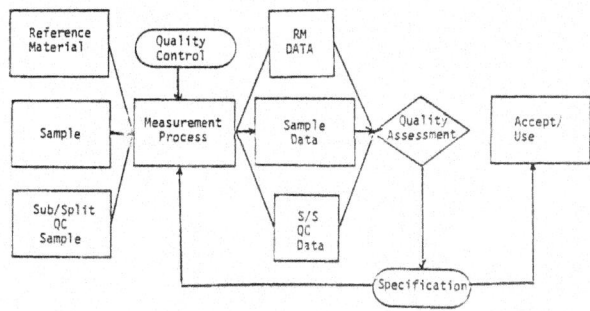

APPROACHES TO QUALITY ASSURANCE

Craftsman/Artisan Approach

    Responsibility - Craftsman
    Effectiveness
        Depends on Craftsman's Knowledge, Skill, Dedication

    Requirements
        Highly Skilled/Dedicated Craftsman
        Atmosphere which Encourages Excellence

    Outstanding Characteristics
        Accuracy, Redundancy

    Works Best for
        Complex Investigations

    Controls
        Peer Review

Formal Quality Assurance Program

    Responibility - Management
    Effectiveness
        Depends on Defined Protocols, Trained Operators,
        Strict Compliance

    Requirements
        Infallible Protocols, Competent Staff

    Outstanding Characteristics
        High Precision, Efficiency

    Works Best For
        Routine, Recurring, Well-Defined Problems

    Controls
        Quality Assurance Program/Office

> "The English Language is the most important scientific instrument at your disposal. Learn to use it with precision."
>
> C. W. Foulk

QUALITY ASSURANCE OF LARGE AND SMALL OPERATIONS

COMPARISONS

LARGE PRODUCTION FACILITY vs. JOB SHOP
MASS PRODUCTION vs. CUSTOM PRODUCTION

LARGE OPERATIONS

SYSTEM CAN BE WELL DEVELOPED
FINE-TUNING JUSTIFIED
COMMON CAUSES IDENTIFIABLE/CONTROLLABLE
SPECIFIC QA POSSIBLE
SPECIAL CAUSES RECOGNIZABLE

SMALL OPERATIONS

EXPEDIENCY INHIBITS SYSTEM DEVELOPMENT
MINIMAL FINE-TUNING JUSTIFIED
COMMON CAUSES/SPECIAL CAUSES INDISTINGUISHABLE
GENERAL QA MORE APPLICABLE

SMALL OPERATIONS

DO NOTHING vs. DO SOMETHING

SPECIFICS

HOT SPOT APPROACH
KEEP RECORDS OF OFFENDERS/PROBLEMS
IDENTIFY HOT SPOTS BY PARETO ANALYSIS

MAGNITUDE
COST/RISK
EASE OF SOLUTION

CONCENTRATION ON HOT SPOTS

| METHODOLOGY | EQUIPMENT |
| MATERIALS | PERSONNNEL |

GENERAL

IDENTIFY NON-SPECIFIC PROBLEMS

INDEPENDENT OF

WHAT ANALYZED?   WHO ANALYZES?
         HOW ANALYZED?

EXAMPLES - COMMON OPERATIONS

CHEMICAL TREATMENTS	REAGENTS/SOLVENTS/WATER
EXTRACTIONS	STANDARDS
CONTAMINATION	HOUSEKEEPING
SAMPLE PREPARATION	FACILITIES
DOCUMENTATION/RECORDS/REPORTS	MAINTENANCE
SUPERVISION	CONTROL CHARTS

GENERAL

CATEGORIZATION APPROACH

FIND COMMONALITIES

MATERIALS	METHODOLOGY
OPERATIONS	OPERATORS

POOL EXPERIENCE

CONTROL ON BASIS OF COMMONALITIES

GLP'S
GMP'S
MINI-SOP'S

QUALITY ASSESSMENT ON BASIS OF COMMONALITIES

TYPICAL REFERENCE MATERIALS
TYPICAL CONTROL CHARTS

II
MEASUREMENT
AS A
PROCESS
CHEMICAL ANALYSIS
AS A
SYSTEM

MEASUREMENT AS A PROCESS

PRODUCT IS NUMBERS

MEASUREMENT PROCESS CAN BE IN CONTROL

STATE OF STATISTICAL CONTROL

RANDOMNESS

ACCURACY AND PRECISION ARE CHARACTERISTICS OF THE PROCESS

CAN BE APPLIED TO THE NUMBERS PRODUCED

THE BODY OF DATA

SUPPORTS INDIVIDUAL MEASUREMENTS

(and vice versa)

PROPERTIES OF MEASUREMENT PROCESSES

Repeated measurements will disagree

Means of repeated measurements will disagree

Measurements at different times, places, by different operators/ apparatus/methods will disagree

Some questions that need answers:

    Within what limits would an additional measurement by the same instrument agree when measuring some stable quantity?
    Would the agreement be poorer if the time interval between repetitions were increased?
    What if different instruments from the same manufacturer were used?
    If two or more types (or manufacturers) were used, how much disagreement would be expected?
    What effect does geometry (orientation, etc.) have on the measurement?
    What about environmental conditions-temperature, moisture, etc.?
    Is the result dependent on the procedure used?
    Do different operators show persistent differences in values?
    Are there instrumental biases or differences due to reference standards or calibrations?

The answers to such questions serve to define the measurement process-the process whose "output" we seek to characterize. Likewise, the inability to answer such questions indicates weakness in knowledge of the measurement processs.

MEASUREMENT as a PROCESS

CHEMICAL ANALYSIS AS A MEASUREMENT SYSTEM

PROCESS

    A progressive action, or a series of acts, especially in the regular course of performing, producing, or making something.

SYSTEM

    An aggregation or assemblage of objects or processes united by some form of regular interaction or interdependence.

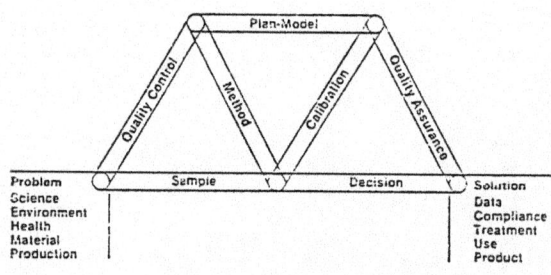

CHEMICAL MEASUREMENT

SITUATIONS

MEASUREMENT VARIANCE SIGNIGICANT -

    Property Variance Not Significant
        Measurement Variance
            Known
            Estimated

MEASUREMENT VARIANCE SIGNIFICANT -

    Property Variance Significant
        Measurement Variance
            Known
            Estimated

        Property Variance
            Estimated

MEASUREMENT VARIANCE NOT SIGNIFICANT -

    Property Variance Significant
    Property Variance
        Estimated

ANALYTICAL PROBLEM SOLVING

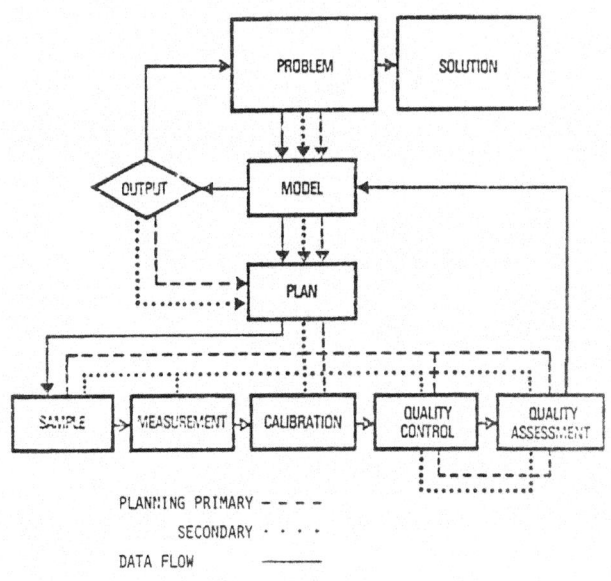

PLANNING PRIMARY - - - -
      SECONDARY · · · ·
DATA FLOW ————

QUESTIONS ANALYTICAL CHEMISTS MUST OFTEN ANSWER

1. Mean of n measurements   III-16

2. Standard deviation of the measurements   III-16, III-17, III-18

3. Confidence interval for mean of n measurements   III-20, III-21

4. Mean of property measured   III-16

5. Standard deviation of property   III-16, III-18

6. How does measured property compare with that from another laboratory
   III-23

7. Is the precision of measurement within specification/expectation
   III-24

8. What confidence do I have in my precision   III-21

9. How do precisions of two laboratories/methods compare   III-24

   Differ? Larger? Smaller?

10. Do measurements indicate product/value is within specification/
    expectations   III-23

    Differ?   Larger?   Smaller?

11. How to combine data from one/several sources to improve

        estimate of mean   III-36
        estimate of variance   III-16

12. How to identify outliers   III-30

13. How to evaluate measurement and property variances when both are or may
    be present   III-18

14. How to estimate number of measurements/samples to

        estimate mean with prescribed precision   V-4
        estimate variance(s) with prescribed precision   III-21

15. Limits for a given percentage of population of measurements/samples –
    statistical tolerance limits   III-22

III

STATISTICAL

CONCEPTS

SELECTED STATISTICAL CONCEPTS
USEFUL WHEN EVALUATING
MEASUREMENTS AND MEASUREMENT PROCESSES

Basic Reference

M. G. Natrella, NBS Handbook 91, Experimental Statistics

QUANTIFICATION - Getting a Number
UNCERTAINTY - Putting ± Limits on the Number
MEASUREMENT AS A PROCESS
  o Product is Numbers
  o Process Stays in Lab, Numbers Go Out
MEASUREMENT PROCESS CAN BE "IN CONTROL"
ACCURACY AND PRECISION ARE CHARACTERISTICS OF THE PROCESS
Can be Applied to the Numbers Produced

STATE OF STATISTICAL CONTROL
Measurements Behave Like Random Samples from a Probability Distribution

POPULATION - All
SAMPLE - Some

STATISTICAL TECHNIQUES
are
TOOLS
Rather Than
ENDS

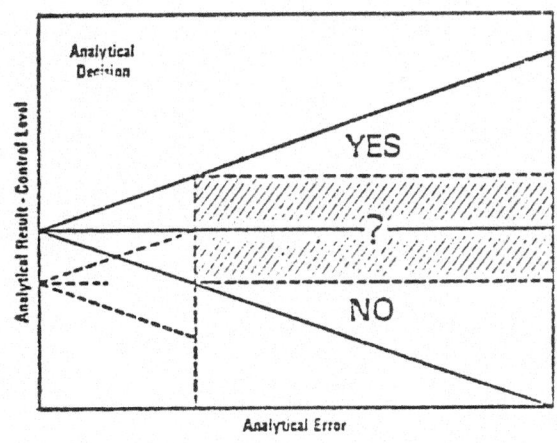

CONSEQUENCES OF ERROR

SOURCES OF UNCERTAINTY

ERRORS

    Systematic
        Known
        Unknown

    Random
        Accidential
        Random Component of Systematic

    Chance Causes
    Assignable Causes

BLUNDERS

AXIOMS

A measurement is . . .
ACCURATE when the value reported does not differ from the <u>true</u> value.
UNBIASED when the error of the limiting mean is zero; freedom from systematic error.
BIASED when the error of the limiting mean is not zero; influenced by systematic error.

COROLLARIES

ERROR in reported values occurs as a result of bias and imprecision.
ACCURATE METHOD - One <u>capable</u> of providing precise and unbiased results (within acceptable <u>limits</u>).
ACCURATE and PRECISE are relative terms

PRECISION AND ACCURACY

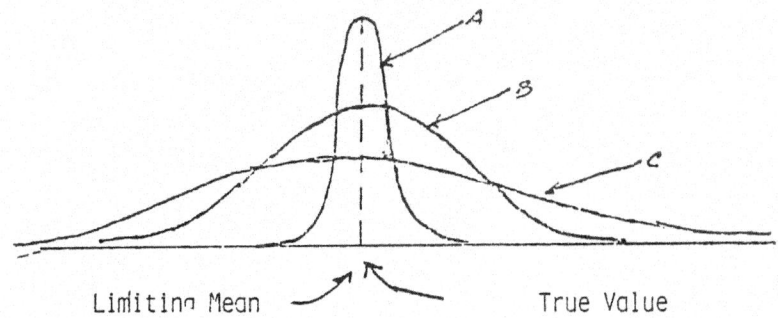

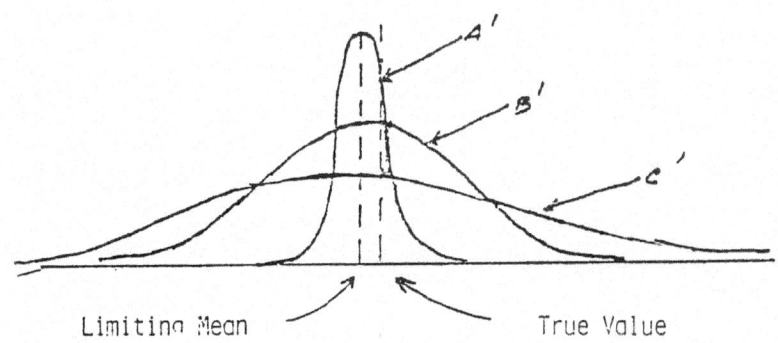

ACCURACY = PRECISION + BIAS

PRECISION - BIAS - ACCURACY

	A	B	C	D
	10	12	9.8	10.5
	14	14	10.1	10.6
	6	9	10.0	10.8
	11	11	9.9	10.6
	13	10	10.1	10.4
	8	15	10.3	10.8
	9	13	9.9	10.4
	7	8	10.3	10.5
	12	16	9.8	10.6
	10	12	10.1	10.5
$\bar{x}$	10.0	12.0	10.03	10.57
s	2.6	2.6	.18	.14
c.l.	±1.9	±1.9	±.13	±.10

TEST SAMPLE VALUE 10.15 ± 0.05

PRECISION IS THE FIRST REQUIREMENT

o Necessary to Define Operational Characteristics
o Good Precision Necessary to Detect Small Constant Errors

".... A measurement process must have attained a state of statistical control; until a measurement operation has been "debugged" to the extent that it has attained a state of statistical control, it cannot be regarded in any logical sense as measuring anything at all."

C. Eisenhart - Ref. 13

"Reproducibility is desirable, but it should not be forgotten that it may be achieved just as easily by insensitivity as by an increase in precision.

Example:

   All men are two meters tall - give or take a meter."

Anon

PROBABILITY PLOTS

RANK DATA according to value

n = total number of data points

i = rank = 1 to n

CALCULATE PROBABILITY PLOT POSITION

$$F_i = \frac{100\ (i-0.5)}{n}$$

PLOT on PROBABILITY PAPER

Scale chosen for distribution postulated
F_i on X axis
Value on Y axis

INTERPRET on BASIS OF FIT TO A STRAIGHT LINE

See Reference 38

"Everybody believes in the exponential law of errors; the experimenters because they think it can be proved by mathematics; and the mathematicians because they believe it has been established by observation."

Lippman, quoted by Poincare

"In applying statistical theory, the main consideration is not what shape the universe is, but whether there is any universe at all. No universe can be assumed nor statistical theory applied unless the observations show statistical control .... Very often, the experimenter instead of rushing in to apply statistics and statistical methods, should be more concerned about attaining statistical control and asking himself whether any predictions at all (the only purpose of his experiment) by statisical theory or otherwise can be made."

W. Edwards Deming
Some Theory of Sampling
J. Wiley, pp. 502-3 (1950)

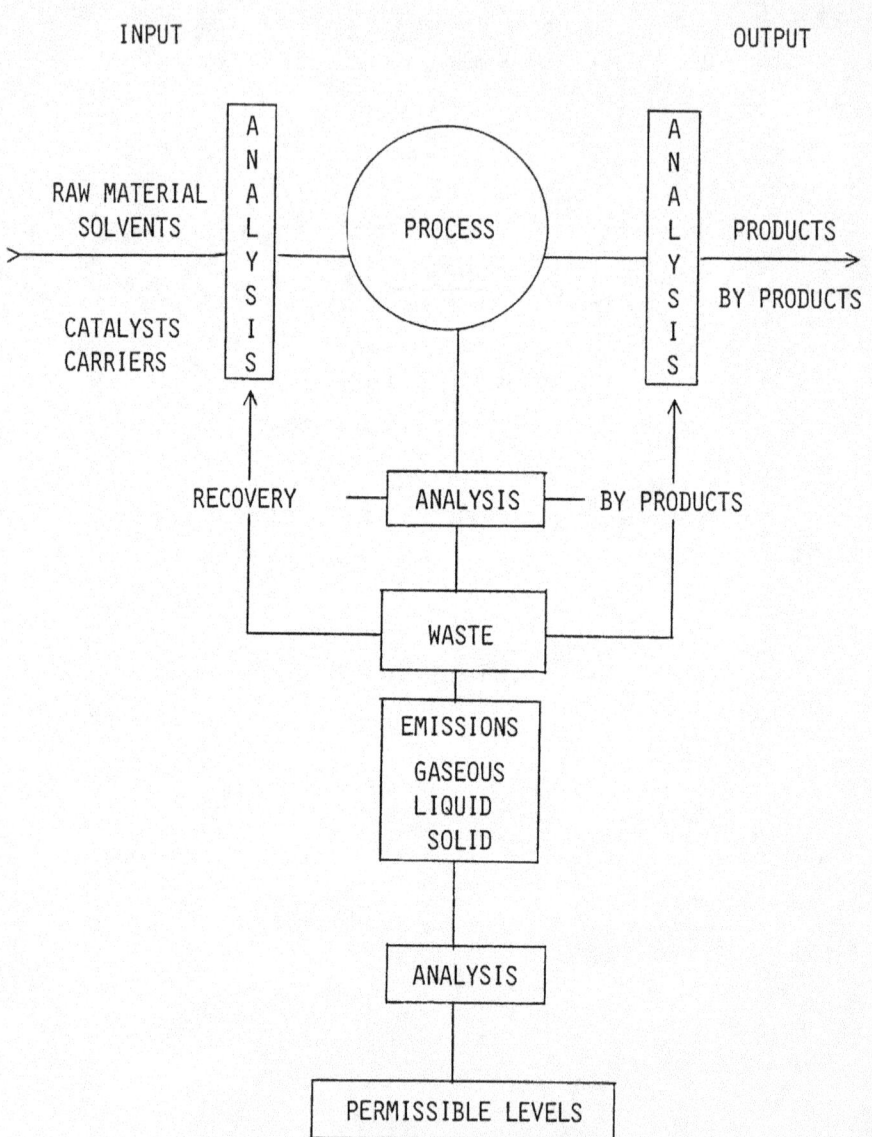

POLLUTION PATHWAYS

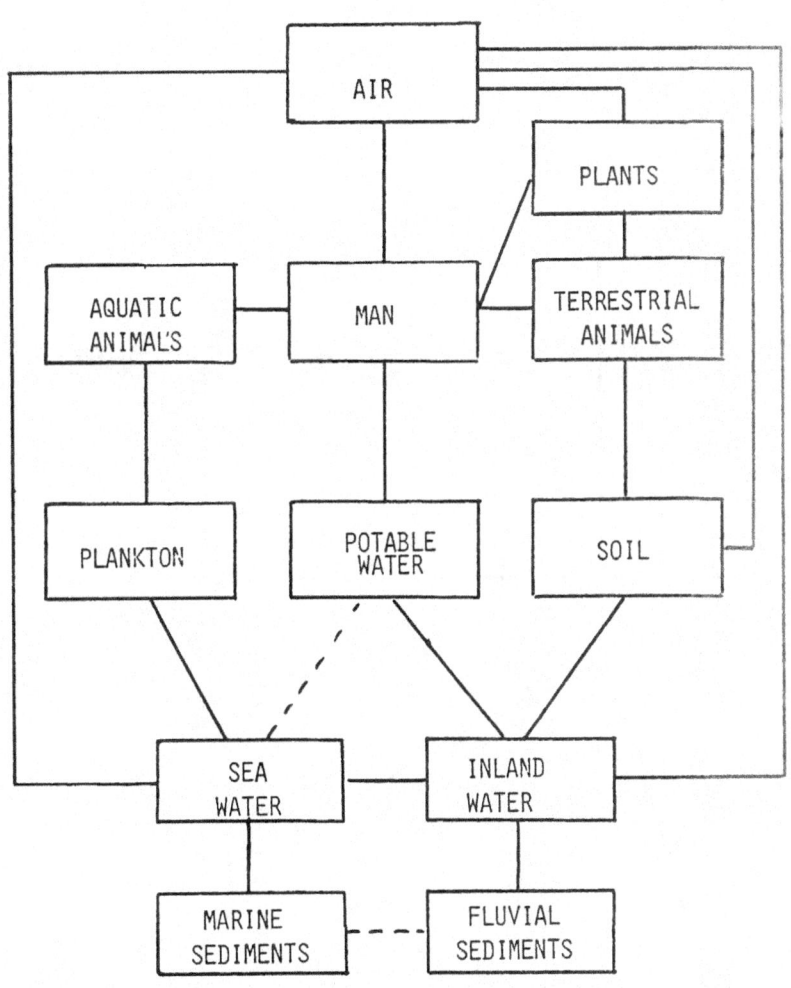

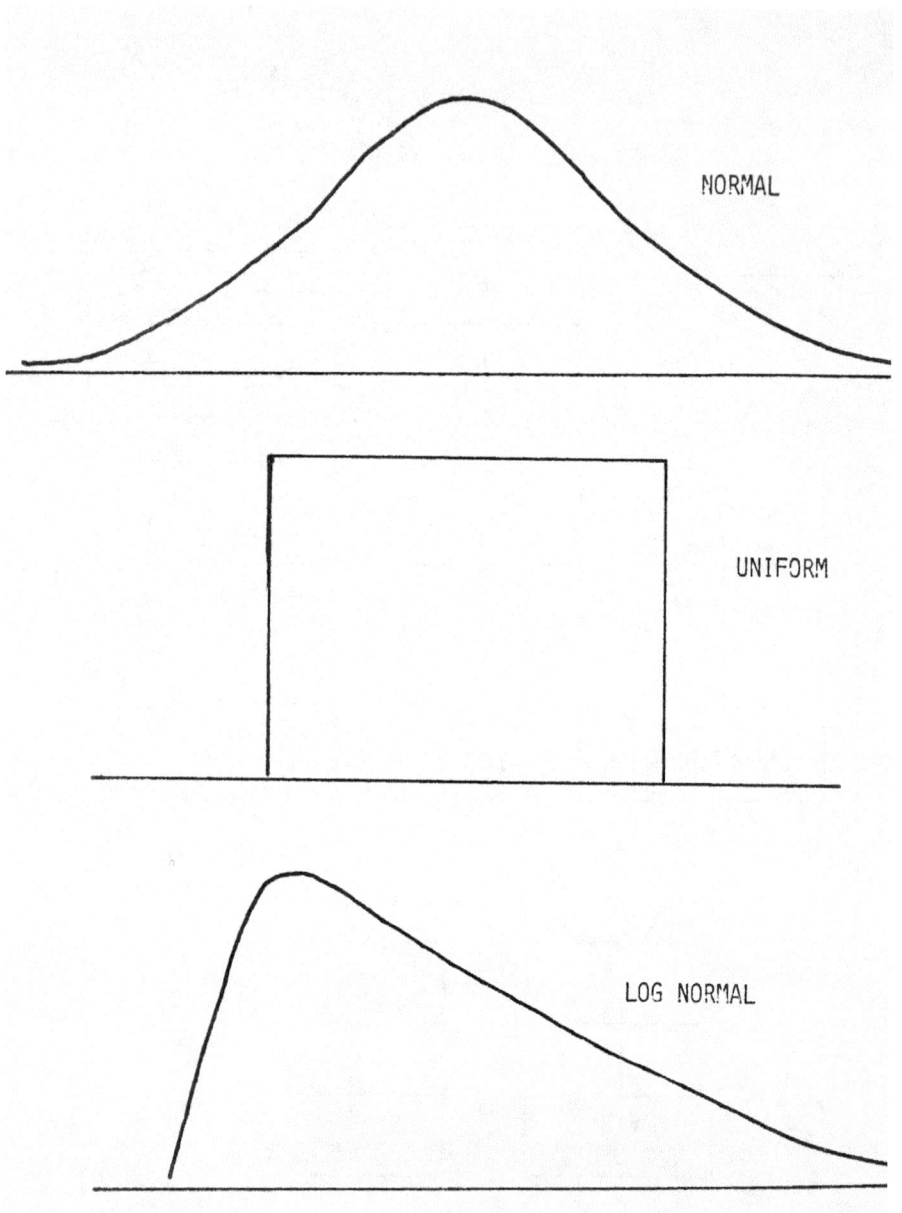

DISTRIBUTIONS

EXAMPLES OF DISTRIBUTIONS

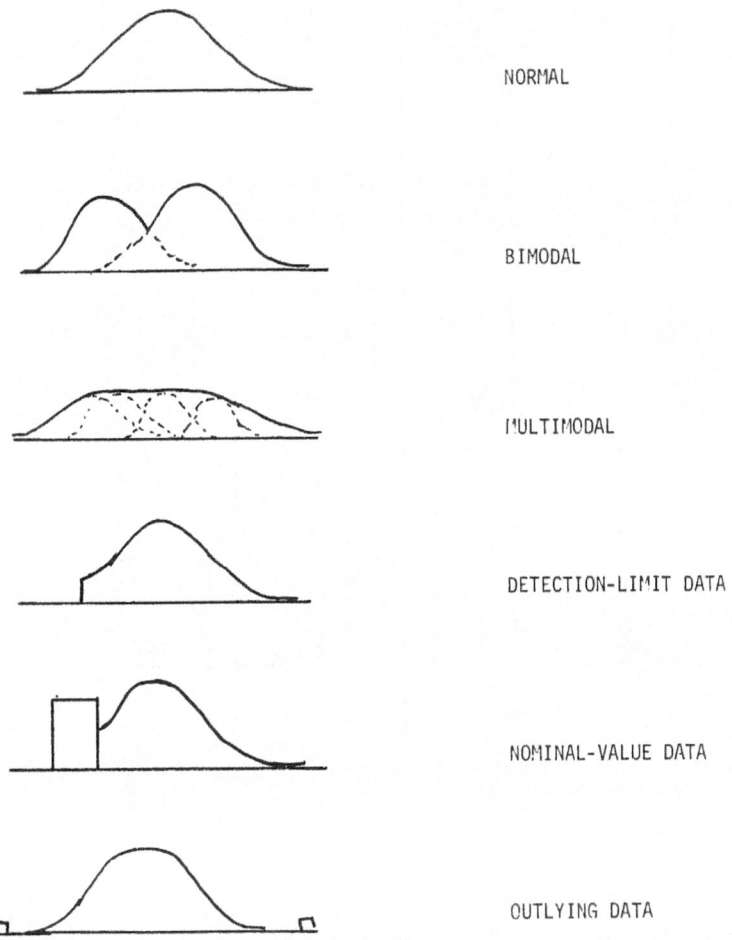

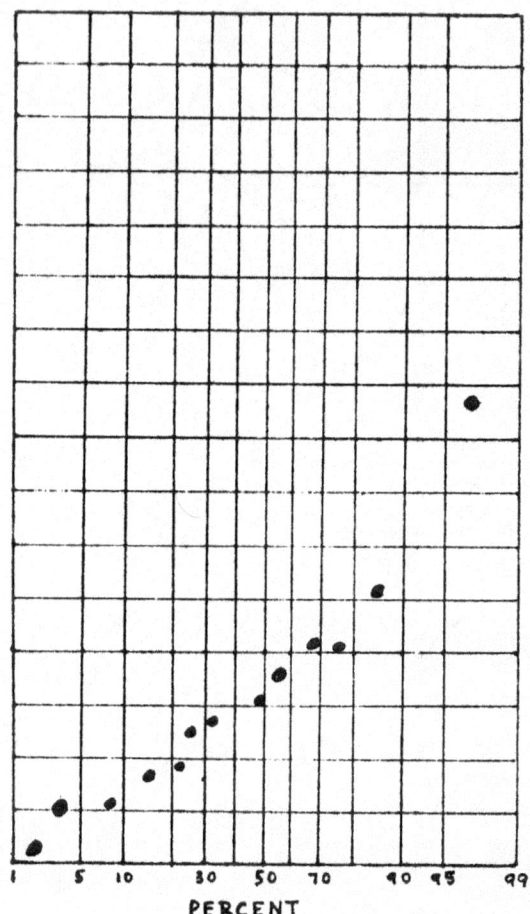

III-10

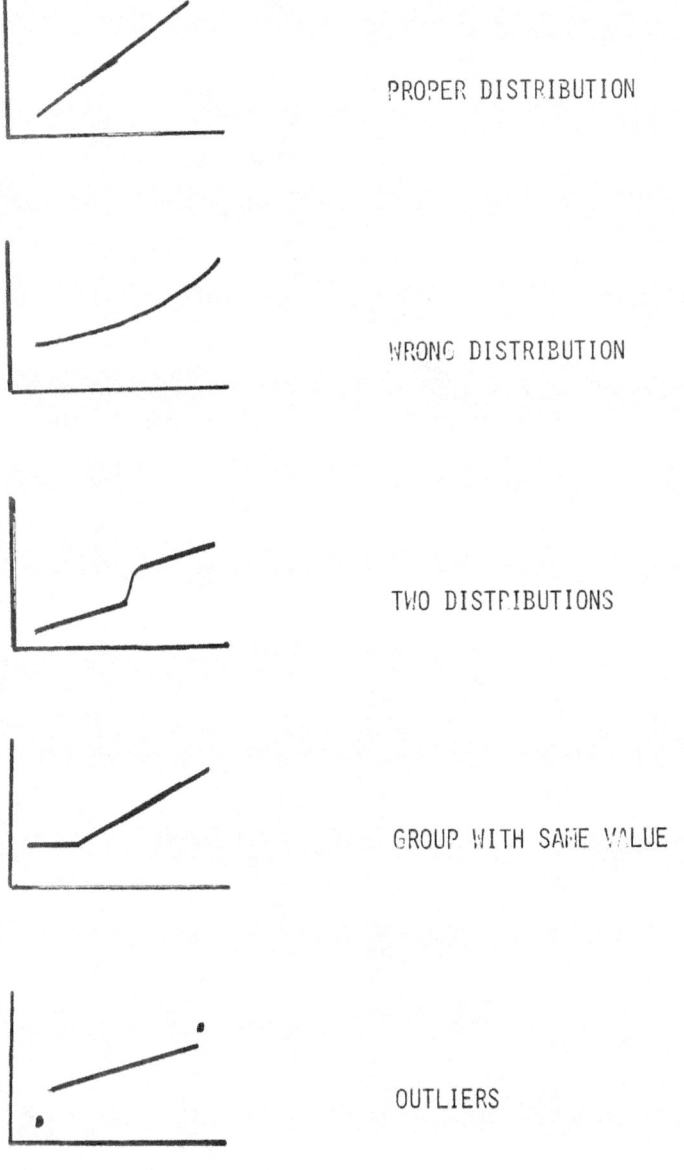

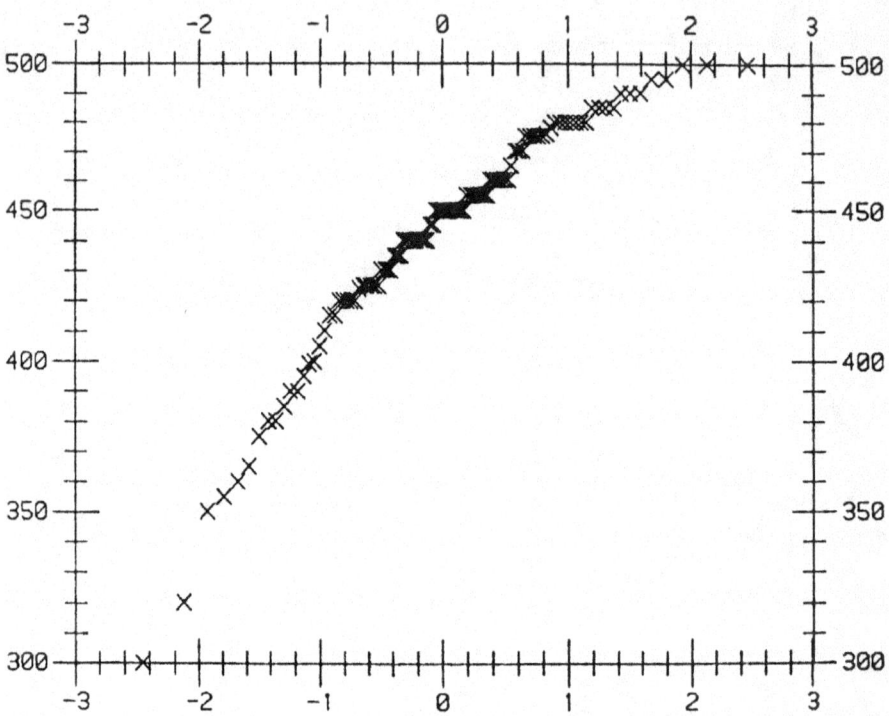

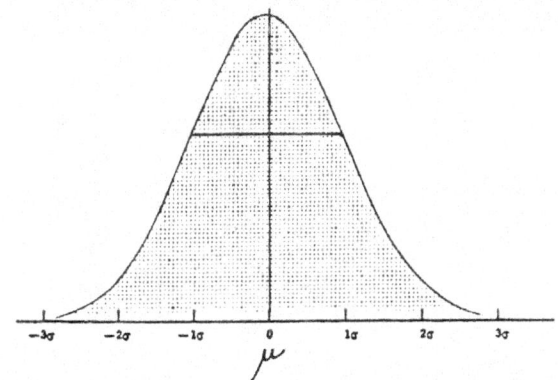

THE
NORMAL
LAW OF ERROR
STANDS OUT IN THE
EXPERIENCE OF MANKIND
AS ONE OF THE BROADEST
GENERALIZATIONS OF NATURAL
PHILOSOPHY ♦ IT SERVES AS THE
GUIDING INSTRUMENT IN RESEARCHES
IN THE PHYSICAL AND SOCIAL SCIENCES AND
IN MEDICINE AGRICULTURE AND ENGINEERING ♦
IT IS AN INDISPENSABLE TOOL FOR THE ANALYSIS AND THE
INTERPRETATION OF THE BASIC DATA OBTAINED BY OBSERVATION AND EXPERIMENT

$$y = \frac{1}{\sigma\sqrt{2\pi}}\, e^{-\frac{(x-\mu)^2}{2\sigma^2}}$$

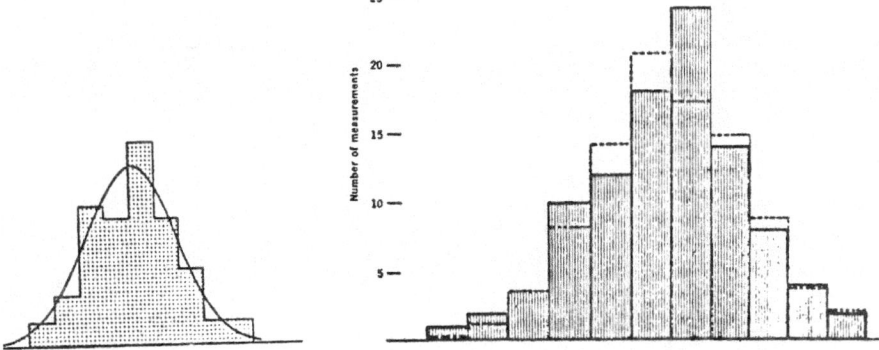

Table 9. Ordinates and areas for the normal curve of error

x values given in multiples, k, of the standard deviation	y values given in multiples of 1/standard deviation	per cent of area included between ordinates at $-k_y$ and $+k_y$
± 0.00	0.3989	0.0000
0.25	0.3867	0.1974
0.50	0.3521	0.3829
0.75	0.3011	0.5467
1.00	0.2420	0.6827
1.25	0.1826	0.7887
1.50	0.1295	0.8664
1.75	0.0863	0.9199
2.00	0.0540	0.9545
2.25	0.0317	0.9756
2.50	0.0175	0.9876
2.75	0.0091	0.9940
3.00	0.0044	0.9973

MULTIPLES OF σ

III-14

BASIC REQUIREMENTS

STABLE

INDEPENDENT

RANDOM

SOME GROSS DEVIATIONS FROM RANDOMNESS

Problem	Diagnostic
TREND	Plots Fits Control Chart
Short-Term Trend	Runs Up and Down * Run Above and Below the Mean* Control Charts
Jumps or Shifts	t-Tests Control Charts

* See e.g., Acheson J. Duncan, Quality Control and Applied Statistics, 4th Edition, Irwin, Homewood, ILL, 1974.

POPULATION VALUES AND SAMPLE ESTIMATES

Population	Sample
All measurements	Sample of n measurements
	$x_1, x_2, \ldots x_n$
Mean m	$\bar{x} = \dfrac{\Sigma X_i}{n}$
Standard Deviation σ	$s = \sqrt{\dfrac{\Sigma(x_i - \bar{x})^2}{n - 1}}$
Variance σ^2	$s^2 = \dfrac{\Sigma(x_i - \bar{x})^2}{n - 1}$

Alternative Computing Formula:

$$s = \sqrt{\dfrac{n\Sigma x^2 - (\Sigma x)^2}{n(n - 1)}}$$

or
PUSH BUTTON

$df = \nu$ = degrees of freedom, often $n - 1$

$s_{\bar{x}} = \dfrac{s}{\sqrt{n}}$ = standard error of mean

$\dfrac{s}{\bar{x}}$ = coefficient of variation

$\dfrac{s}{\bar{x}} \times 100$ = Relative standard deviation

POOLING ESTIMATES OF VARIANCE
GENERAL CASE

$$S_p^2 = \dfrac{\nu_1 S_1^2 + \nu_2 S_2^2 + \cdots + \nu_k S_k^2}{\nu_1 + \nu_2 + \cdots + \nu_k}$$

where $\nu_k = n_k - 1$ $df = \Sigma \nu_i$

$$S_p^2 = \dfrac{(n_1-1) S_1^2 + (n_2-1) S_2^2 + \cdots + (n_k-1) S_k^2}{n_1 + n_2 + \cdots + n_k - k}$$

$df = \Sigma n_i - k$

DUPLICATE MEASUREMENTS

$$S_p^2 = \frac{1}{2k} \sum_1^k d_i^2$$

where k = number of sets of duplicates
 d_i = difference of duplicate measurement
 S_p has k degrees of freedom

USE OF RANGE TO ESTIMATE VARIABILITY

Number of Samples k		Size of Sample				
		2	3	4	5	6
1	d_2^*	1.41	1.91	2.24	2.48	2.67
	ν	1.00	1.98	2.93	3.83	4.68
3	d_2^*	1.23	1.77	2.12	2.38	2.58
	ν	2.83	5.86	8.44	11.1	13.6
5	d_2^*	1.19	1.74	2.10	2.36	2.56
	ν	4.59	9.31	13.9	18.4	22.6
10	d_2^*	1.16	1.72	2.08	2.34	2.55
	ν	8.99	18.4	27.6	36.5	44.9
20	d_2^*	1.14	1.70	2.07	2.33	2.54
	ν	17.8	36.5	55.0	72.7	89.6
∞	d_2	1.13	1.69	2.06	2.33	2.53

$$\bar{R}/d_2^* \to \sigma$$

Adapted from Lloyd S. Nelson, J. Qual. Tech. $\underline{7}$ No. 1, January 1975.

WITHIN AND BETWEEN STANDARD DEVIATION

I. GENERAL CASE

n Replicates of k samples (e.g., homogeneity)
n Replicates on h occasions (e.g., long-/short-term s)
n Replicates by k laboratories (e.g., Methodology)
where $2 < n < 5$ $k > 6$

Tabulate

R_1 $\bar{x}_1$
R_2 $\bar{x}_2$
:   :   R = Highest-Lowest
:   :
R_k $\bar{x}_k$

1. Calculate $\bar{R} = \dfrac{R_1 + R_2 + \cdots + R_k}{k}$

2. Calculate $S_W = \bar{R}/d_2^*$ (d_2^* for appropriate k)
   ν = see Table III-17

3. Calculate $S_{\bar{x}}$

   $\bar{\bar{x}} = \dfrac{\bar{x}_1 + \bar{x}_2 + \cdots \bar{x}_k}{k}$ $S_{\bar{x}} = \sqrt{\dfrac{\Sigma(\bar{x}_i - \bar{\bar{x}})^2}{k-1}}$

4. Calculate $S_B = \sqrt{(S_{\bar{x}}^2 - \dfrac{S_W^2}{n})}$

   $\nu = k - 1$

II. SPECIAL CASE - S_W known

   $n > 1$

   Omit steps 1 and 2

TABLE 2-3. PROPAGATION OF ERROR FORMULAS FOR SOME SIMPLE FUNCTIONS

(X and Y are assumed to be independent.)

Function form	Approximate formula for $s_{\bar{w}}^2$
$m_w = Am_x + Bm_y$	$A^2 s_{\bar{x}}^2 + B^2 s_{\bar{y}}^2$
$m_w = \dfrac{m_x}{m_y}$	$\left(\dfrac{\bar{x}}{\bar{y}}\right)^2 \left(\dfrac{s_{\bar{x}}^2}{\bar{x}^2} + \dfrac{s_{\bar{y}}^2}{\bar{y}^2}\right)$
$m_w = \dfrac{1}{m_y}$	$\dfrac{s_{\bar{y}}^2}{\bar{y}^4}$
$m_w = \dfrac{m_x}{m_x + m_y}$	$\left(\dfrac{\bar{w}}{\bar{x}}\right)^4 (\bar{y}^2 s_{\bar{x}}^2 + \bar{x}^2 s_{\bar{y}}^2)$
$m_w = \dfrac{m_x}{1 + m_x}$	$\dfrac{s_{\bar{x}}^2}{(1 + \bar{x})^4}$
*$m_w = m_x m_y$	$(\bar{x}\bar{y})^2 \left(\dfrac{s_{\bar{x}}^2}{\bar{x}^2} + \dfrac{s_{\bar{y}}^2}{\bar{y}^2}\right)$
*$m_w = m_x^2$	$4\bar{x}^2 s_{\bar{x}}^2$
$m_w = \sqrt{m_x}$	$\dfrac{1}{4}\dfrac{s_{\bar{x}}^2}{\bar{x}}$
*$m_w = \ln m_x$	$\dfrac{s_{\bar{x}}^2}{\bar{x}^2}$
*$m_w = k m_x^a m_y^b$	$\bar{w}^2 \left(a^2 \dfrac{s_{\bar{x}}^2}{\bar{x}^2} + b^2 \dfrac{s_{\bar{y}}^2}{\bar{y}^2}\right)$
*$m_w = e^{m_x}$	$e^{2\bar{x}} s_{\bar{x}}^2$
$W = 100 \dfrac{s_x}{\bar{x}}$ (=coefficient of variation)	$\dfrac{\bar{w}^2}{2(n-1)}$ (not directly derived from the formulas)†

*Distribution of $\bar{w}$ is highly skewed and normal approximation could be seriously in error for small n.

†See, for example, *Statistical Theory with Engineering Applications*, by A. Hald, John Wiley & Sons, Inc., New York, 1952, p. 301.

WHAT TO REPORT AND WHY

1. Report at Least:

    $\bar{x}$
    s
    n

2. Why?

    How good $\bar{x}$ is as an estimate of m depends on s amd n. How good s is as an estimate of variability depends on n, more precisely on degrees of freedom associated with s. (In a simple series of measurements, degrees of freedom = n - 1).

3. Other

    Confidence Interval
    Tolerance Interval

CONFIDENCE INTERVAL FOR THE MEAN

(σ Known)

Form:

$$\bar{x} \pm z \frac{\sigma}{\sqrt{n}}$$

Where:

   $\bar{x}$ = sample mean
   n = no. of measurements in sample
   z depends only on the "confidence level"

For example,

   For a 90% conf. int., z = 1.645
         95% conf. int., z = 1.960 ⟶ 2
         99% conf. int., z = 2.576
         99.8% conf. int., z = 3.09 ⟶ 3

See Chapter 2, Handbook 91, also III-36.

CONFIDENCE INTERVAL FOR THE MEAN

(σ Not Known, Using s From Sample)

Form:

$$\bar{x} \pm t_c \frac{s}{\sqrt{n}}$$

Where:

$\bar{x}$ = sample mean
n = no. of measurements in sample
s = calculated from sample
t_c from t-table (e.g., Table pIII-38) depends on:
confidence level and degrees of freedom associated with $\underline{s}$
(D.F. = n - 1 in simple series).

For example, if $\underline{s}$ has 4. D.F. (n = 5),

For 90% conf. int., t_c = 2.132
95% conf. int., t_c = 2.776
99% conf. int., t_c = 4.604

See Chapter 2, Handbook 91.

CONFIDENCE INTERVAL FOR σ

Lower limit: $B_L s$

Upper Limit: $B_U s$
Interval: $B_L s$ to $B_U s$
B_L and B_U from table (e.g., Table pIII-39) depend on
confidence level and degrees of freedom associated with $\underline{s}$.

For example, if s has 4 D. F.,

95% conf. int., .599 s to 2.567 s
99% conf. int., .4865 s to 3.892 s

Another example, if s has 30 D.F.,

95% conf. int., .791 s to 1.321 s
99% conf. int., .740 s to 1.457 s

STATISTICAL TOLERANCE INTERVALS

(m and σ Known)

Form:

$m \pm z\sigma$

where:

z depends only on P, the percentage of the individual measurements to be included in the interval.

For example,

To include 50%, z = .67
90%, z = 1.645
95%, z = 1.96 ⟶ 2
99%, z = 2.576
99.8%, z = 3.09 ⟶ 3

See Chapter 2, Handbook 91.

STATISTICAL TOLERANCE INTERVALS

(Using $\bar{x}$ and s from Sample)

Form:
$\bar{x} \pm k s$

k depends on three things;
    P, the proportion or percentage of individuals to be included;
    Y, the confidence coefficient to be associated with the interval;
    D.F., the degrees of freedom associated with <u>s</u>.

	For P = 90% Y = .90	For P = 99% Y = .99
n = 2 (D.F. = 1)	k = 15.98	k = 242.
n = 5 (D.F. = 4)	k = 3.494	k = 10.26
n = 10 (D.F. = 9)	k = 2.535	k = 5.59

See Chapter 2, Handbook 91, See also III-41.

REMINDER

Confidence Interval: is expected to include the mean or the standard deviation.
Tolerance Interval: is expected to include P% of the individual measurements.
The Interpretation of "Expected" is the Same in Each Case

COMPARING A SET OF MEASUREMENTS WITH A STANDARD MEAN VALUE (m_o)

Given: m_o

Set of Measurements: $x_1, x_2 \cdots x_n$

1. Calculate: $\bar{x}$, s (or use σ known)
2. Calculate: confidence interval for mean

$$\bar{x} \pm z \frac{\sigma}{\sqrt{n}} \text{ or } \bar{x} \pm \frac{ts}{\sqrt{n}}$$

3. m_o is given.

    If confidence interval includes m_o, set is consistent with standard mean value, m_o.

See Chapter 3, Handbook 91.

COMPARING TWO SETS WITH REGARD TO THEIR MEANS

(Using Confidence Intervals)

Have: two sets of measurements, A and B

Calculate:
$\bar{x}_A$ $\bar{x}_B$
s_A s_B
n_A n_B
(df = $n_A - 1$)   (df = $n_B - 1$)

Confidence Interval for Difference of Means:

$$(\bar{x}_A - \bar{x}_B) \pm t_c\, s_p \sqrt{\frac{n_A + n_B}{n_A\, n_B}}$$

where t_c depends on:
- confidence level, D.F. for S_p
- D.F. for $s_p = n_A + n_B - 2$ here
- from Table (e.g., A-4 in Handbook 91), or page III-38.

and

$$s_p = \sqrt{\frac{(n_A - 1)s^2_A + (n_B - 1)s^2_B}{n_A + n_B - 2}}$$

If confidence interval includes zero, consider sets consistent with regard to mean.

See Chapter 3, Handbook 91.

Note: If S_A and S_B are considered to be different, calculate individual confidence intervals and check for overlap.

COMPARING A SET OF MEASUREMENTS WITH A STANDARD VALUE FOR VARIABILITY (σ_o)

Given: σ_o

Set of Measurements: $x_1, x_2, \cdots x_n$

1. Calculate: s

2. Calculate: confidence interval for σ
   B_{LS} to B_{US}

3. Given σ_o.
   If confidence interval includes σ_o, set is consistent with this standard value.

See Chapter 4, Handbook 91.

COMPARING TWO SETS WITH REGARD TO VARIABILITY

(F Test)

Have: Set A   Set B

$$s^2_A \qquad\qquad s^2_B$$
$$n_A \qquad\qquad n_B$$
$$(D.F. = n_A - 1) \qquad (D.F. = n_B - 1)$$

Calculate: $F = \dfrac{s^2_A}{s^2_B}$

Compare with table value of F (e.g., Table A-5, Handbook 91) or p. III-

depends on: - significance level of test
                  - degrees of freedom of numerator and denominator

See Chapter 4, Handbook 91; see also III-40.

IS LINEAR ASSUMPTION JUSTIFIED

Grapical Approach

    Plot data and attempt to draw "best" straight line
If straight line is "easy" to draw, linearity is probably justified

    Fit function, $y = \beta_0 + \beta_1 x$, by least squares

        Compute

            β_0 and s_{β_0}
            β_1 and s_{β_0}

        Judge significance of β_0 nd β_1

Compute $\Delta y = y_{obs} - y_{calc}$.

    Tabulate sign changes and sign follows
Should not be large differences between them
Examine lengths of runs
Plot Δ vs. x and look for absence of functional trends.

CORRELATION COEFFICIENT APPROACH COMPUTE CORRELATION COEFFICIENT, r

$$r = \frac{n \Sigma xy - \Sigma x \Sigma y}{\sqrt{[n\Sigma x^2 - (\Sigma x)^2][n\Sigma y^2 - (\Sigma y)^2]}}$$

$r = 0$, no correlation

$r = +1$, perfect positive correlation

$r = -1$, perfect negative correlation

$r =$ immediate values, usual case, use Table A17 to interpret, see p III-29.

Note: Correlation coefficient is often misused. See H. T. Arm, "The Significance of the Correlation Coefficient for Analyzing Engineering Data," Materials Research and Standards Vol. 11, No. 5, pp. 16-19 (1971).

LINEAR RELATIONSHIPS

TABLE 5-1. SUMMARY OF FOUR CASES OF LINEAR RELATIONSHIPS

	Functional (F)		Statistical (S)	
	FI	FII	SI	SII
Distinctive Features and Example	x and y are linearly related by a mathematical formula, $y = \beta_0 + \beta_1 x$, or $z = \beta_0' + \beta_1' y$, which is not observed exactly because of disturbances or errors in one or both variables. Example: Determination of elastic constant of a spring which obeys Hooke's law. x = accurately-known weight applied, Y = measured value of corresponding elongation y.		X = Height Y = Weight Both measured on a random sample of individuals. X is *not* selected but "comes with" sample unit.	X = Height (preselected values) Y = Weight of individuals of preselected height X is measured beforehand; only *selected* values of X are used at which to measure Y.
Errors of Measurement	Measurement error affects Y only.	X and Y both subject to error.	Ordinarily negligible compared to variation among individuals.	Same as in SI.
Form of Line Fitted	$Y = b_0 + b_1 x$	See Paragraph 5-4.3.	$\hat{Y}_x = b_0 + b_1 X$ $\hat{X}_y = b_0' + b_1' Y$	$\hat{Y}_x = b_0 + b_1 X$ only.
Procedure for Fitting	See Paragraphs 5-4.2, and basic worksheet.	Procedure depends on what assumptions can be made. See Paragraph 5-4.3.	See Paragraph 5-5.1 and basic worksheet.	See Paragraph 5-5.2 and basic worksheet.
Correlation Coefficient	Not applicable	Not applicable	Sample estimate is $r = \dfrac{S_{xy}}{\sqrt{S_{xx}} \cdot \sqrt{S_{yy}}}$ See Paragraph 5-5.1.5.	Correlation may exist in the population, but r computed from *such* an experiment would provide a distorted estimate of the correlation.

Chapter 5, Handbook 91.

BASIC WORKSHEET FOR ALL TYPES OF LINEAR RELATIONSHIPS

X denotes _____ Y denotes _____

$\Sigma X =$ _____ $\Sigma Y =$ _____

$\bar{X} =$ _____ $\bar{Y} =$ _____

Number of points: $n =$ _____

Step (1) ΣXY = _____

(2) $(\Sigma X)(\Sigma Y)/n$ = _____

(3) S_{xy} = Step (1) − Step (2)

(4) ΣX^2 = _____   (7) ΣY^2 = _____

(5) $(\Sigma X)^2/n$ = _____   (8) $(\Sigma Y)^2/n$ = _____

(6) S_{xx} = Step (4) − Step (5)   (9) S_{yy} = Step (7) − Step (8)

(10) $b_1 = \dfrac{S_{xy}}{S_{xx}}$ = Step (3) ÷ Step (6)   (14) $\dfrac{(S_{xy})^2}{S_{xx}}$ = _____

(11) $\bar{Y}$ = _____   (15) $(n-2)s_Y^2$ = Step (9) − Step (14)

(12) $b_1\bar{X}$ = _____   (16) s_Y^2 = Step (15) ÷ (n − 2)

(13) $b_0 = \bar{Y} - b_1\bar{X}$ = Step (11) − Step (12) s_Y = _____

> Equation of the line:
> $Y = b_0 + b_1 X$
>
> _____
>
> $s_{b_1} =$ _____
>
> $s_{b_0} =$ _____

Estimated variance of the slope:

$s_{b_1}^2 = \dfrac{s_Y^2}{S_{xx}}$ = Step (16) ÷ Step (6)

Estimated variance of intercept:

$s_{b_0}^2 = s_Y^2 \left\{\dfrac{1}{n} + \dfrac{\bar{X}^2}{S_{xx}}\right\}$ = _____

Note: The following are algebraically identical:
$S_{xx} = \Sigma(X - \bar{X})^2$; $S_{yy} = \Sigma(Y - \bar{Y})^2$; $S_{xy} = \Sigma(X - \bar{X})(Y - \bar{Y})$.

Chapter 5, Handbook 91.

TABLE 5-4. SOME LINEARIZING TRANSFORMATIONS

If the Relationship Is of the Form:	Plot the Transformed Variables		Fit the Straight Line $Y_T = b_0 + b_1 X_T$	Convert Straight Line Constants (b_0 and b_1) To Original Constants:	
	$Y_T =$	$X_T =$		$b_0 =$	$b_1 =$
$Y = a + \dfrac{b}{X}$	Y	$\dfrac{1}{X}$	Use the procedures of Paragraph 5-4.1.1.	a	b
$Y = \dfrac{1}{a + bX}$, or $\dfrac{1}{Y} = a + bX$	$\dfrac{1}{Y}$	X	In all formulas given there, substitute values of Y_T for Y and values of X_T for X, as appropriate.	a	b
$Y = \dfrac{X}{a + bX}$	$\dfrac{X}{Y}$	X		a	b
$Y = ab^X$	$\log Y$	X		$\log a$	$\log b$
$Y = ae^{bX}$	$\log Y$	X		$\log a$	$b \log e$
$Y = aX^b$	$\log Y$	$\log X$		$\log a$	b
$Y = a + bX^n$, where n is known	Y	X^n		a	b

Chapter 5, Handbook 91.

TABLE A-17. CONFIDENCE BELTS FOR THE CORRELATION COEFFICIENT
(CONFIDENCE COEFFICIENT .95)

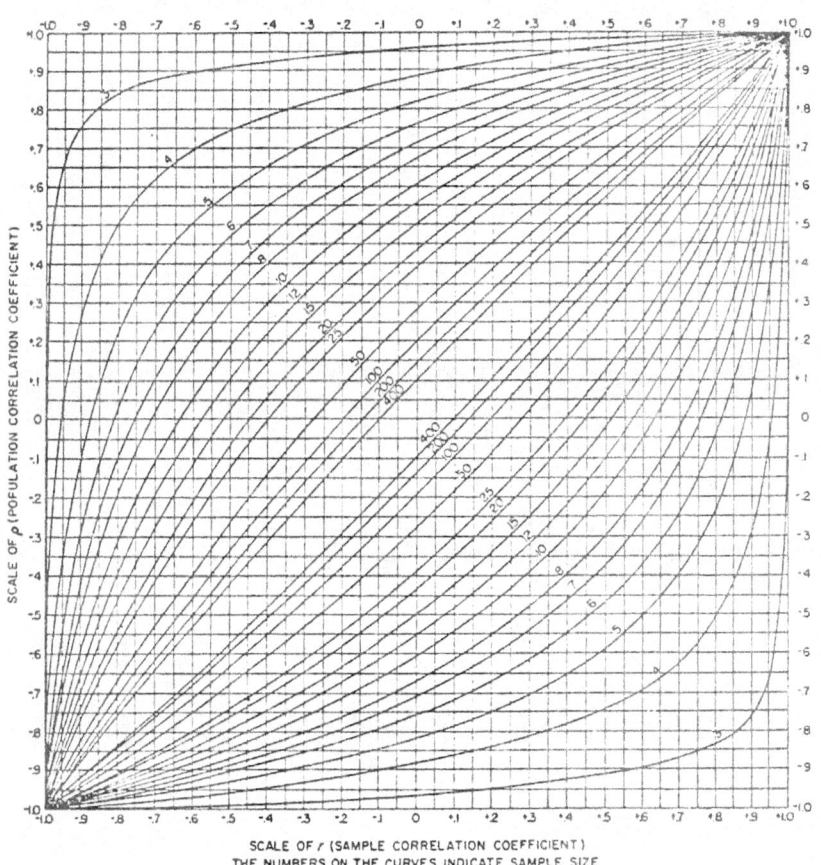

Section 5, Handbook 91.

OUTLIERS

Outlier - An intruder/foreigner in an otherwise well-defined population
Why Worry About Outliers?

   Presence of Outliers

   o   Clouds conclusions
   o   May nullify statistical conclusions
   o   May introduce error in

       o   theoretical treatment
       o   theoretical conclusions drawn from data

   o   May indicate trouble and indicate need to further investigate
       and/or trouble shoot

   Cause for Alert

Identification of Outliers

   Plotting Results
   Residual Plots
   Ranking Results
   Control Chart Results
   Empirical Tests
   Statistical Tests

Rejection of <u>true</u> outliers is not only merited but recommended/condoned
But this <u>cannot be arbitrary</u> and/or <u>capricious</u>
Guidance for Rejection of Data

   Rejection for Assignable Cause

          o   Result of inspection
          o   Result of audit
          o   Out-of-Control usually a set of data will be involved

   Action -

          o   Document and reject
          o   Identify cause
          o   Keep record of causes
          o   Take corrective actions

Too many outliers spells TROUBLE

Rejection for Unassignable Cause

    Example
    Blunders

        Wrong Sample
        Transcription Error
        Transposition Error
        Misreadings

    Malfunction
    Unsuspected Loss
    Contamination

Must distingush between true outliers and extreme random variations

Empirical Test-Huge Error

    If Outlier is Suspected

        Analyze Data Excluding Suspect, Compute Standard Deviation
        From Set or Curve
        Determine Deviation of Suspect from Mean
        or Deviation from Curve
        Compare Difference with Standard Deviation
        Reject Suspect If

$$\frac{\text{DIFFERENCE}}{S} > 4$$

PROBLEM OF REJECTING OBSERVATIONS

17-3.1 WHEN EXTREME OBSERVATIONS IN EITHER DIRECTION ARE CONSIDERED REJECTABLE

17-3 1.1 Population Mean and Standard Deviation Unknown — Sample in Hand is the Only Source of Information.

(The Dixon Criterion)

Procedure

Rank Data in Order of Increasing Size

$X_1 \ X_2 \ X_3 \ X_4 \ \ldots \ X_{n-1}, \ X_n$

Choose α, the probability or risk we are willing to take of rejecting an observation that really belongs in the group.

If:
$3 \leq n \leq 7$	Compute r_{10}
$8 \leq n \leq 10$	Compute r_{11}
$11 \leq n \leq 13$	Compute r_{21}
$14 \leq n \leq 25$	Compute r_{22}

where r_{ij} is computed as follows:

r_{ij}	If X_n is Suspect	If X_1 is Suspect
r_{10}	$(X_n - X_{n-1})/(X_n - X_1)$	$(X_2 - X_1)/(X_n - X_1)$
r_{11}	$(X_n - X_{n-1})/(X_n - X_2)$	$(X_2 - X_1)/(X_{n-1} - X_1)$
r_{21}	$(X_n - X_{n-2})/(X_n - X_2)$	$(X_3 - X_1)/(X_{n-1} - X_1)$
r_{22}	$(X_n - X_{n-2})/(X_n - X_3)$	$(X_3 - X_1)/(X_{n-2} - X_1)$

Look up $r_{1-\alpha}$ for the r_{ij} from Step (2), in Table A-14.

If $r_{ij} > r_{1-\alpha}$, reject the suspect observation; otherwise, retain it.

TABLE A-14. CRITERIA FOR REJECTION OF OUTLYING OBSERVATIONS

Statistic	Number of Observations, n	Upper Percentiles						
		.70	.80	.90	.95	.98	.99	.995
r_{10}	3	.684	.781	.886	.941	.976	.988	.994
	4	.471	.560	.679	.765	.846	.889	.926
	5	.373	.451	.557	.642	.729	.780	.821
	6	.318	.386	.482	.560	.644	.698	.740
	7	.281	.344	.434	.507	.586	.637	.680
r_{11}	8	.318	.385	.479	.554	.631	.683	.725
	9	.288	.352	.441	.512	.587	.635	.677
	10	.265	.325	.409	.477	.551	.597	.639
r_{21}	11	.391	.442	.517	.576	.638	.679	.713
	12	.370	.419	.490	.546	.605	.642	.675
	13	.351	.399	.467	.521	.578	.615	.649
r_{22}	14	.370	.421	.492	.546	.602	.641	.674
	15	.353	.402	.472	.525	.579	.616	.647
	16	.338	.386	.454	.507	.559	.595	.624
	17	.325	.373	.438	.490	.542	.577	.605
	18	.314	.361	.424	.475	.527	.561	.589
	19	.304	.350	.412	.462	.514	.547	.575
	20	.295	.340	.401	.450	.502	.535	.562
	21	.287	.331	.391	.440	.491	.524	.551
	22	.280	.323	.382	.430	.481	.514	.541
	23	.274	.316	.374	.421	.472	.505	.532
	24	.268	.310	.367	.413	.464	.497	.524
	25	.262	.304	.360	.406	.457	.489	.516

RANKING TEST TO IDENTIFY OUTLIERS

Table 4. Approximate 5% two-tail limits for ranking scores

No. of Labs.	Number of Materials											
	3	4	5	6	7	8	9	10	11	12	13	14	15
3		4 12	5 15	7 17	8 20	10 22	12 24	13 27	15 29	17 31	19 33	20 36	22 38
4		4 16	6 19	8 22	10 25	12 28	14 31	16 34	18 37	20 40	22 43	24 46	26 49
5			5 19	7 23	9 27	11 31	13 35	16 38	18 42	21 45	23 49	26 52	28 56	31 59
6	3 18	5 23	7 28	10 32	12 37	15 41	18 45	21 49	23 54	26 58	29 62	32 66	35 70
7	3 21	5 27	8 32	11 37	14 42	17 47	20 52	23 57	26 62	29 67	32 72	36 76	39 81
8	3 24	6 30	9 36	12 42	15 48	18 54	22 59	25 65	29 70	32 76	36 81	39 87	43 92
9	3 27	6 34	9 41	13 47	16 54	20 60	24 66	27 73	31 79	35 85	39 91	43 97	47 103
10	4 29	7 37	10 45	14 52	17 60	21 67	26 73	30 80	34 87	38 94	43 100	47 107	51 114
11	4 32	7 41	11 49	15 57	19 65	23 73	27 81	32 88	36 96	41 103	46 110	51 117	55 125
12	4 35	7 45	11 54	15 63	20 71	24 80	29 88	34 96	39 104	44 112	49 120	54 128	59 136
13	4 38	8 48	12 58	16 68	21 77	26 86	31 95	36 104	42 112	47 121	52 130	58 138	63 147
14	4 41	8 52	12 63	17 73	22 83	27 93	33 102	38 112	44 121	50 130	56 139	61 149	67 158
15	4 44	8 56	13 67	18 78	23 89	29 99	35 109	41 119	47 129	53 139	59 149	65 159	71 169

W. J. Youden, in NBS Special Publication 300, H. Ku, Editor, Ref. 26.

COMPUTING MEAN OF DATA SETS

CASE I   All Sets Have Same Precision
Case IA Same Number of Measurements in Each Set

$$\bar{\bar{x}} = \frac{\bar{x}_1 + \bar{x}_2 + \cdots \bar{x}_k}{k}$$

$$\sigma_{\bar{\bar{x}}} = \frac{\sigma_{\bar{x}}}{\sqrt{k}} = \frac{\sigma_x}{\sqrt{nk}}$$

$$s_{\bar{\bar{x}}} = \frac{s_p}{\sqrt{nk}}$$

n = no. of meas. in each set.

Case IB Unequal Number of Measurements in Each Set
Assign wts., W, equivalent to no. of meas. in set

$$\bar{\bar{x}} = \frac{\bar{x}_1 W_1 + \bar{x}_2 W_2 + \cdots \bar{x}_k W_k}{W_1 + W_2 \cdots W_k}$$

Case II Data Sets Have Different Precisions Due to Number and Imprecision

Compute Wts., $W = \dfrac{1}{\sigma_{\bar{x}}^2}$

$$\sigma_{\bar{x}} = \frac{\sigma_x}{\sqrt{n}}$$

$$s_{\bar{x}} = \frac{s_x}{\sqrt{n}}$$

$$\bar{\bar{X}} = \frac{\bar{X}_1 \cdot \frac{1}{\sigma_{\bar{X}_1}^2} + \bar{X}_2 \cdot \frac{1}{\sigma_{\bar{X}_2}^2} + \cdots \bar{X}_k \cdot \frac{1}{\sigma_{\bar{X}_k}^2}}{\frac{1}{\sigma_{\bar{X}_1}^2} + \frac{1}{\sigma_{\bar{X}_2}^2} + \cdots \frac{1}{\sigma_{\bar{X}_k}^2}}$$

Ordinarily, s will be known, rather than σ
$\bar{\bar{X}}$ ≈ above with s substituted for σ

$$\sigma_{\bar{\bar{X}}}^2 = \frac{1}{W_1 + W_2 + \cdots W_k}$$

where $W = \frac{1}{\sigma_{\bar{\bar{X}}}^2}$ $W \approx \frac{1}{s_{\bar{X}}^2}$

Z-factors for 2-sided confidence interval

Confidence Level	Z Factor
50%	0.67
67	1.00
75	1.15
90	1.645
95	1.960
95.28	2.000
99.00	2.575
99.74	3
99.9934	4
99.99995	5
10^{-9}	6
10^{-12}	7
10^{-15}	8
$10^{-18.9}$	9
10^{-23}	10

Combining Data Sets

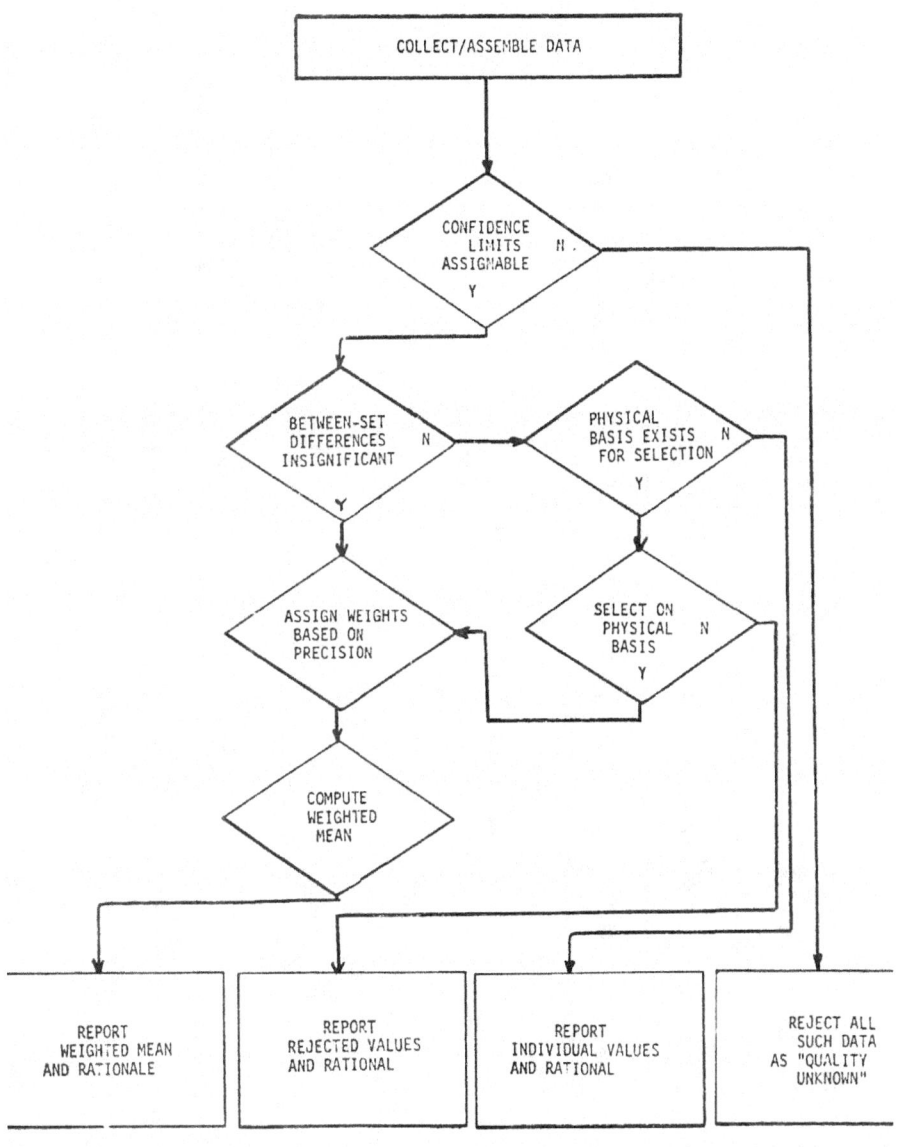

TABLE A-4. PERCENTILES OF THE t DISTRIBUTION

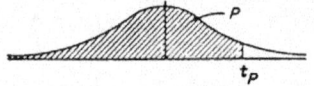

df	$t_{.60}$	$t_{.70}$	$t_{.80}$	$t_{.90}$	$t_{.95}$	$t_{.975}$	$t_{.99}$	$t_{.995}$
1	.325	.727	1.376	3.078	6.314	12.706	31.821	63.657
2	.289	.617	1.061	1.886	2.920	4.303	6.965	9.925
3	.277	.584	.978	1.638	2.353	3.182	4.541	5.841
4	.271	.569	.941	1.533	2.132	2.776	3.747	4.604
5	.267	.559	.920	1.476	2.015	2.571	3.365	4.032
6	.265	.553	.906	1.440	1.943	2.447	3.143	3.707
7	.263	.549	.896	1.415	1.895	2.365	2.998	3.499
8	.262	.546	.889	1.397	1.860	2.306	2.896	3.355
9	.261	.543	.883	1.383	1.833	2.262	2.821	3.250
10	.260	.542	.879	1.372	1.812	2.228	2.764	3.169
11	.260	.540	.876	1.363	1.796	2.201	2.718	3.106
12	.259	.539	.873	1.356	1.782	2.179	2.681	3.055
13	.259	.538	.870	1.350	1.771	2.160	2.650	3.012
14	.258	.537	.868	1.345	1.761	2.145	2.624	2.977
15	.258	.536	.866	1.341	1.753	2.131	2.602	2.947
16	.258	.535	.865	1.337	1.746	2.120	2.583	2.921
17	.257	.534	.863	1.333	1.740	2.110	2.567	2.898
18	.257	.534	.862	1.330	1.734	2.101	2.552	2.878
19	.257	.533	.861	1.328	1.729	2.093	2.539	2.861
20	.257	.533	.860	1.325	1.725	2.086	2.528	2.845
21	.257	.532	.859	1.323	1.721	2.080	2.518	2.831
22	.256	.532	.858	1.321	1.717	2.074	2.508	2.819
23	.256	.532	.858	1.319	1.714	2.069	2.500	2.807
24	.256	.531	.857	1.318	1.711	2.064	2.492	2.797
25	.256	.531	.856	1.316	1.708	2.060	2.485	2.787
26	.256	.531	.856	1.315	1.706	2.056	2.479	2.779
27	.256	.531	.855	1.314	1.703	2.052	2.473	2.771
28	.256	.530	.855	1.313	1.701	2.048	2.467	2.763
29	.256	.530	.854	1.311	1.699	2.045	2.462	2.756
30	.256	.530	.854	1.310	1.697	2.042	2.457	2.750
40	.255	.529	.851	1.303	1.684	2.021	2.423	2.704
60	.254	.527	.848	1.296	1.671	2.000	2.390	2.660
120	.254	.526	.845	1.289	1.658	1.980	2.358	2.617
∞	.253	.524	.842	1.282	1.645	1.960	2.326	2.576

TABLE A-20. FACTORS FOR COMPUTING TWO-SIDED CONFIDENCE LIMITS FOR σ

Degrees of Freedom r	$\alpha = .05$		$\alpha = .01$		$\alpha = .001$	
	B_U	B_L	B_U	B_L	B_U	B_L
1	17.79	.3576	86.31	.2969	844.4	.2480
2	4.859	.4581	10.70	.3879	33.29	.3291
3	3.183	.5178	5.449	.4453	11.65	.3824
4	2.567	.5590	3.892	.4865	6.938	.4218
5	2.248	.5899	3.175	.5182	5.085	.4529
6	2.052	.6143	2.764	.5437	4.128	.4784
7	1.918	.6344	2.498	.5650	3.551	.5000
8	1.820	.6513	2.311	.5830	3.167	.5186
9	1.746	.6657	2.173	.5987	2.894	.5348
10	1.686	.6784	2.065	.6125	2.689	.5492
11	1.638	.6896	1.980	.6248	2.530	.5621
12	1.598	.6995	1.909	.6358	2.402	.5738
13	1.564	.7084	1.851	.6458	2.298	.5845
14	1.534	.7166	1.801	.6549	2.210	.5942
15	1.509	.7240	1.758	.6632	2.136	.6032
16	1.486	.7308	1.721	.6710	2.073	.6116
17	1.466	.7372	1.688	.6781	2.017	.6193
18	1.448	.7430	1.658	.6848	1.968	.6266
19	1.432	.7484	1.632	.6909	1.925	.6333
20	1.417	.7535	1.609	.6968	1.886	.6397
21	1.404	.7582	1.587	.7022	1.851	.6457
22	1.391	.7627	1.568	.7074	1.820	.6514
23	1.380	.7669	1.550	.7122	1.791	.6568
24	1.370	.7709	1.533	.7169	1.765	.6619
25	1.360	.7747	1.518	.7212	1.741	.6668
26	1.351	.7783	1.504	.7253	1.719	.6713
27	1.343	.7817	1.491	.7293	1.698	.6758
28	1.335	.7849	1.479	.7331	1.679	.6800
29	1.327	.7880	1.467	.7367	1.661	.6841
30	1.321	.7909	1.457	.7401	1.645	.6880
31	1.314	.7937	1.447	.7434	1.629	.6917
32	1.308	.7964	1.437	.7467	1.615	.6953
33	1.302	.7990	1.428	.7497	1.601	.6987
34	1.296	.8015	1.420	.7526	1.588	.7020
35	1.291	.8039	1.412	.7554	1.576	.7052
36	1.286	.8062	1.404	.7582	1.564	.7083
37	1.281	.8085	1.397	.7608	1.553	.7113
38	1.277	.8106	1.390	.7633	1.543	.7141
39	1.272	.8126	1.383	.7658	1.533	.7169
40	1.268	.8146	1.377	.7681	1.523	.7197
41	1.264	.8166	1.371	.7705	1.515	.7223
42	1.260	.8184	1.365	.7727	1.506	.7248
43	1.257	.8202	1.360	.7748	1.498	.7273
44	1.253	.8220	1.355	.7769	1.490	.7297
45	1.249	.8237	1.349	.7789	1.482	.7320
46	1.246	.8253	1.345	.7809	1.475	.7342
47	1.243	.8269	1.340	.7828	1.468	.7364
48	1.240	.8285	1.335	.7847	1.462	.7386
49	1.237	.8300	1.331	.7864	1.455	.7407
50	1.234	.8314	1.327	.7882	1.449	.7427

TABLE A-5 (Continued). PERCENTILES OF THE F DISTRIBUTION

$$F_{.975}(n_1, n_2)$$

n_1 = degrees of freedom for numerator

n_2 = degrees of freedom for denominator

n_2 \ n_1	1	2	3	4	5	6	7	8	9	10	12	15	20	24	30	40	60	120	∞
1	647.8	799.5	864.2	899.6	921.8	937.1	948.2	956.7	963.3	968.6	976.7	984.9	993.1	997.2	1001	1006	1010	1014	1018
2	38.51	39.00	39.17	39.25	39.30	39.33	39.36	39.37	39.39	39.40	39.41	39.43	39.45	39.46	39.46	39.47	39.48	39.49	39.50
3	17.44	16.04	15.44	15.10	14.88	14.73	14.62	14.54	14.47	14.42	14.34	14.25	14.17	14.12	14.08	14.04	13.99	13.95	13.90
4	12.22	10.65	9.98	9.60	9.36	9.20	9.07	8.98	8.90	8.84	8.75	8.66	8.56	8.51	8.46	8.41	8.36	8.31	8.26
5	10.01	8.43	7.76	7.39	7.15	6.98	6.85	6.76	6.68	6.62	6.52	6.43	6.33	6.28	6.23	6.18	6.12	6.07	6.02
6	8.81	7.26	6.60	6.23	5.99	5.82	5.70	5.60	5.52	5.46	5.37	5.27	5.17	5.12	5.07	5.01	4.96	4.90	4.85
7	8.07	6.54	5.89	5.52	5.29	5.12	4.99	4.90	4.82	4.76	4.67	4.57	4.47	4.42	4.36	4.31	4.25	4.20	4.14
8	7.57	6.06	5.42	5.05	4.82	4.65	4.53	4.43	4.36	4.30	4.20	4.10	4.00	3.95	3.89	3.84	3.78	3.73	3.67
9	7.21	5.71	5.08	4.72	4.48	4.32	4.20	4.10	4.03	3.96	3.87	3.77	3.67	3.61	3.56	3.51	3.45	3.39	3.33
10	6.94	5.46	4.83	4.47	4.24	4.07	3.95	3.85	3.78	3.72	3.62	3.52	3.42	3.37	3.31	3.26	3.20	3.14	3.08
11	6.72	5.26	4.63	4.28	4.04	3.88	3.76	3.66	3.59	3.53	3.43	3.33	3.23	3.17	3.12	3.06	3.00	2.94	2.88
12	6.55	5.10	4.47	4.12	3.89	3.73	3.61	3.51	3.44	3.37	3.28	3.18	3.07	3.02	2.96	2.91	2.85	2.79	2.72
13	6.41	4.97	4.35	4.00	3.77	3.60	3.48	3.39	3.31	3.25	3.15	3.05	2.95	2.89	2.84	2.78	2.72	2.66	2.60
14	6.30	4.86	4.24	3.89	3.66	3.50	3.38	3.29	3.21	3.15	3.05	2.95	2.84	2.79	2.73	2.67	2.61	2.55	2.49
15	6.20	4.77	4.15	3.80	3.58	3.41	3.29	3.20	3.12	3.06	2.96	2.86	2.76	2.70	2.64	2.59	2.52	2.46	2.40
16	6.12	4.69	4.08	3.73	3.50	3.34	3.22	3.12	3.05	2.99	2.89	2.79	2.68	2.63	2.57	2.51	2.45	2.38	2.32
17	6.04	4.62	4.01	3.66	3.44	3.28	3.16	3.06	2.98	2.92	2.82	2.72	2.62	2.56	2.50	2.44	2.38	2.32	2.25
18	5.98	4.56	3.95	3.61	3.38	3.22	3.10	3.01	2.93	2.87	2.77	2.67	2.56	2.50	2.44	2.38	2.32	2.26	2.19
19	5.92	4.51	3.90	3.56	3.33	3.17	3.05	2.96	2.88	2.82	2.72	2.62	2.51	2.45	2.39	2.33	2.27	2.20	2.13
20	5.87	4.46	3.86	3.51	3.29	3.13	3.01	2.91	2.84	2.77	2.68	2.57	2.46	2.41	2.35	2.29	2.22	2.16	2.09
21	5.83	4.42	3.82	3.48	3.25	3.09	2.97	2.87	2.80	2.73	2.64	2.53	2.42	2.37	2.31	2.25	2.18	2.11	2.04
22	5.79	4.38	3.78	3.44	3.22	3.05	2.93	2.84	2.76	2.70	2.60	2.50	2.39	2.33	2.27	2.21	2.14	2.08	2.00
23	5.75	4.35	3.75	3.41	3.18	3.02	2.90	2.81	2.73	2.67	2.57	2.47	2.36	2.30	2.24	2.18	2.11	2.04	1.97
24	5.72	4.32	3.72	3.38	3.15	2.99	2.87	2.78	2.70	2.64	2.54	2.44	2.33	2.27	2.21	2.15	2.08	2.01	1.94
25	5.69	4.29	3.69	3.35	3.13	2.97	2.85	2.75	2.68	2.61	2.51	2.41	2.30	2.24	2.18	2.12	2.05	1.98	1.91
26	5.66	4.27	3.67	3.33	3.10	2.94	2.82	2.73	2.65	2.59	2.49	2.39	2.28	2.22	2.16	2.09	2.03	1.95	1.88
27	5.63	4.24	3.65	3.31	3.08	2.92	2.80	2.71	2.63	2.57	2.47	2.36	2.25	2.19	2.13	2.07	2.00	1.93	1.85
28	5.61	4.22	3.63	3.29	3.06	2.90	2.78	2.69	2.61	2.55	2.45	2.34	2.23	2.17	2.11	2.05	1.98	1.91	1.83
29	5.59	4.20	3.61	3.27	3.04	2.88	2.76	2.67	2.59	2.53	2.43	2.32	2.21	2.15	2.09	2.03	1.96	1.89	1.81
30	5.57	4.18	3.59	3.25	3.03	2.87	2.75	2.65	2.57	2.51	2.41	2.31	2.20	2.14	2.07	2.01	1.94	1.87	1.79
40	5.42	4.05	3.46	3.13	2.90	2.74	2.62	2.53	2.45	2.39	2.29	2.18	2.07	2.01	1.94	1.88	1.80	1.72	1.64
60	5.29	3.93	3.34	3.01	2.79	2.63	2.51	2.41	2.33	2.27	2.17	2.06	1.94	1.88	1.82	1.74	1.67	1.58	1.48
120	5.15	3.80	3.23	2.89	2.67	2.52	2.39	2.30	2.22	2.16	2.05	1.94	1.82	1.76	1.69	1.61	1.53	1.43	1.31
∞	5.02	3.69	3.12	2.79	2.57	2.41	2.29	2.19	2.11	2.05	1.94	1.83	1.71	1.64	1.57	1.48	1.39	1.27	1.00

TABLE A-6 (Continued). FACTORS FOR TWO-SIDED TOLERANCE LIMITS FOR NORMAL DISTRIBUTIONS

n \ P	γ = 0.95					γ = 0.99				
	0.75	0.90	0.95	0.99	0.999	0.75	0.90	0.95	0.99	0.999
2	22.858	32.019	37.674	48.430	60.573	114.363	160.193	188.491	242.300	303.054
3	5.922	8.380	9.916	12.861	16.208	13.378	18.930	22.401	29.055	36.616
4	3.779	5.369	6.370	8.299	10.502	6.614	9.398	11.150	14.527	18.383
5	3.002	4.275	5.079	6.634	8.415	4.643	6.612	7.855	10.260	13.015
6	2.604	3.712	4.414	5.775	7.337	3.743	5.337	6.345	8.301	10.548
7	2.361	3.369	4.007	5.248	6.676	3.233	4.613	5.488	7.187	9.142
8	2.197	3.136	3.732	4.891	6.226	2.905	4.147	4.936	6.468	8.234
9	2.078	2.967	3.532	4.631	5.899	2.677	3.822	4.550	5.966	7.600
10	1.987	2.839	3.379	4.433	5.649	2.508	3.582	4.265	5.594	7.129
11	1.916	2.737	3.259	4.277	5.452	2.378	3.397	4.045	5.308	6.766
12	1.858	2.655	3.162	4.150	5.291	2.274	3.250	3.870	5.079	6.477
13	1.810	2.587	3.081	4.044	5.158	2.190	3.130	3.727	4.893	6.240
14	1.770	2.529	3.012	3.955	5.045	2.120	3.029	3.608	4.737	6.043
15	1.735	2.480	2.954	3.878	4.949	2.060	2.945	3.507	4.605	5.876
16	1.705	2.437	2.903	3.812	4.865	2.009	2.872	3.421	4.492	5.732
17	1.679	2.400	2.858	3.754	4.791	1.965	2.808	3.345	4.393	5.607
18	1.655	2.366	2.819	3.702	4.725	1.926	2.753	3.279	4.307	5.497
19	1.635	2.337	2.784	3.656	4.667	1.891	2.703	3.221	4.230	5.399
20	1.616	2.310	2.752	3.615	4.614	1.860	2.659	3.168	4.161	5.312
21	1.599	2.286	2.723	3.577	4.567	1.833	2.620	3.121	4.100	5.234
22	1.584	2.264	2.697	3.543	4.523	1.808	2.584	3.078	4.044	5.163
23	1.570	2.244	2.673	3.512	4.484	1.785	2.551	3.040	3.993	5.098
24	1.557	2.225	2.651	3.483	4.447	1.764	2.522	3.004	3.947	5.039
25	1.545	2.208	2.631	3.457	4.413	1.745	2.494	2.972	3.904	4.985
26	1.534	2.193	2.612	3.432	4.382	1.727	2.469	2.941	3.865	4.935
27	1.523	2.178	2.595	3.409	4.353	1.711	2.446	2.914	3.828	4.888

TABLE A-36. SHORT TABLE OF RANDOM NUMBERS

```
46 96 85 77 27 92 86 26 45 21 89 91 71 42 64 64 58 22 75 81 74 91 48 46 18
44 19 15 32 63 55 57 77 33 29 45 00 31 34 84 05 72 90 44 27 78 22 07 62 17
34 39 80 62 24 33 81 67 28 11 34 79 26 35 34 23 09 94 00 80 55 31 63 27 91
74 97 80 30 65 07 71 30 01 84 47 45 89 70 74 13 04 90 51 27 61 34 64 87 44
22 14 61 60 85 38 33 71 13 33 72 08 16 13 50 56 48 51 29 43 30 93 45 66 29

40 03 96 40 03 47 24 60 09 21 21 18 00 05 86 52 85 40 73 73 57 68 36 33 91
52 33 76 44 56 15 47 75 78 73 78 19 87 06 98 47 48 02 62 03 42 05 32 55 02
37 59 20 40 93 17 82 24 19 90 60 87 32 74 59 84 24 49 79 17 23 75 83 42 00
11 02 55 57 48 84 74 36 22 67 19 20 15 92 53 37 13 75 54 89 56 73 23 39 07
10 33 79 26 34 54 71 33 89 74 68 48 23 17 49 18 81 05 52 85 70 05 73 11 17

67 59 28 25 47 89 11 65 65 20 42 23 96 41 64 20 30 89 87 64 37 93 36 96 35
93 50 75 20 09 18 54 34 63 02 54 87 23 05 43 36 93 29 97 93 87 08 20 92 93
24 43 23 72 80 64 34 27 23 46 15 36 10 63 21 59 69 76 02 62 31 62 47 60 24
39 91 63 18 38 27 10 78 88 84 42 32 00 97 92 00 04 94 50 05 75 82 70 80 35
74 62 19 67 54 18 28 92 33 69 98 96 74 35 72 11 68 25 08 95 31 79 11 79 54

91 03 35 60 81 16 61 97 25 14 78 21 22 05 25 47 26 37 80 39 19 06 41 02 00
42 57 66 76 72 91 03 63 48 46 44 01 33 53 62 28 80 59 55 05 02 16 13 17 54
06 35 63 06 15 03 72 38 01 53 25 37 06 48 56 19 56 41 29 28 76 49 74 39 50
92 70 96 70 89 80 87 14 25 49 25 94 62 78 26 15 41 39 48 75 64 69 61 06 38
91 08 88 53 52 13 04 82 23 00 26 36 47 44 04 08 84 80 07 44 76 51 52 41 59

68 85 97 74 47 53 90 05 90 84 87 48 25 01 11 05 45 11 43 15 60 40 31 84 59
59 54 13 09 13 80 42 29 63 03 24 64 12 43 28 10 01 65 62 07 79 83 05 59 61
39 18 32 69 53 46 56 19 34 03 60 28 97 31 02 65 47 47 70 39 74 17 30 22 65
67 43 81 09 12 60 19 57 63 78 11 80 10 57 15 70 04 89 81 78 54 84 87 83 42
61 75 37 19 56 90 75 39 03 56 49 92 72 95 27 52 87 47 12 52 54 62 43 23 13

78 10 91 11 00 63 19 63 74 58 69 03 51 38 60 36 53 56 77 06 69 03 89 91 24
93 23 71 58 09 78 08 03 07 71 79 32 25 19 61 04 40 33 12 06 78 91 97 88 95
37 55 48 82 63 89 92 59 14 72 19 17 22 51 90 20 03 64 96 60 48 01 95 44 84
62 13 11 71 17 23 29 25 13 85 33 35 07 60 25 68 57 92 57 11 84 44 01 23 66
29 89 97 47 00 13 20 86 22 45 59 98 64 53 89 64 94 81 65 87 73 81 58 46 42

16 94 85 82 89 07 17 30 29 89 89 80 98 36 25 36 53 02 49 14 34 03 52 09 20
01 93 10 59 75 12 98 84 60 93 68 16 87 60 11 50 46 56 58 45 88 72 50 46 11
95 71 43 68 97 18 85 17 13 08 00 50 77 50 46 92 45 26 97 21 48 22 23 08 32
86 05 33 14 35 48 68 18 36 57 09 62 40 28 87 08 74 79 91 08 27 12 43 32 03
59 30 60 10 41 31 00 69 63 77 01 89 94 60 19 02 70 88 72 33 38 88 20 60 83

05 45 85 40 54 03 98 96 76 27 77 84 80 03 64 60 44 34 54 24 85 20 85 77 32
71 85 17 74 66 27 85 19 55 56 51 35 48 92 32 44 40 47 10 38 22 52 42 29 96
80 20 32 80 98 00 40 92 57 51 52 83 14 55 31 99 73 23 40 07 64 54 44 99 21
13 50 78 02 73 39 66 82 01 28 67 51 75 66 33 97 47 58 42 44 83 09 28 58 06
67 92 65 41 45 36 77 96 46 21 14 39 56 36 70 15 74 43 62 69 32 30 77 28 77

72 56 73 44 26 04 62 81 15 35 79 26 99 57 28 22 25 91 80 62 95 48 98 23 86
28 86 85 64 94 11 58 78 45 56 34 45 91 38 51 10 63 36 87 81 16 77 30 15 36
69 57 40 80 44 94 60 82 94 93 98 01 48 50 57 69 60 77 69 60 74 22 05 77 17
71 20 03 30 79 25 74 17 78 34 54 45 04 77 42 59 75 78 64 99 37 03 18 03 36
89 93 55 98 22 45 12 49 82 71 57 33 28 69 50 59 15 09 25 79 39 42 84 18 70

58 74 82 81 14 02 01 05 77 94 65 57 70 39 42 48 56 84 31 59 18 70 41 74 60
50 54 73 81 91 07 81 26 25 45 49 61 22 88 41 20 00 15 59 93 51 60 65 65 63
49 33 72 90 10 20 65 28 44 63 95 88 75 78 69 24 41 65 36 10 84 10 32 00 93
11 85 01 43 65 02 85 69 56 88 34 29 64 35 48 16 70 11 77 83 01 34 82 91 04
34 22 46 41 84 74 27 02 57 77 47 93 72 02 95 63 75 74 69 69 61 34 31 92 13
```

IV

MODELING

MODELING

More or Less Idealized Representation of an Often Complex Reality

One Can Model

    Objects
    Phenomena
    Processes

Qualities of a Satisfactory Model

- Undistorted, Unbiased picture (A group of models may be needed)
- Simplicity (as far as possible) to allow interpretation (sets of simple logical elements may be best)
- Coverage of all essential parts (no significant gaps)
- Interlinkage defined for all related components

EXAMPLES OF MODELS

- Automobile Emissions

    Emissions During Driving Cycle
    Simulated Cycle
    Composite Sample
    Relation to Other Models

- Air Dispersion Model

    Stack Concentration
    Dispersed Concentration
        Plume Considerations
        Fumigations

- Dissolved Matter

- TSP

- Respirable Fraction

- Biological Availability

- BOD - TOC - TOD

- Volatile Organics

- Flammability

A RANKING OF ENVIRONMENTAL DATA IN TERMS OF THE POLLUTION CONTROL PROCESS

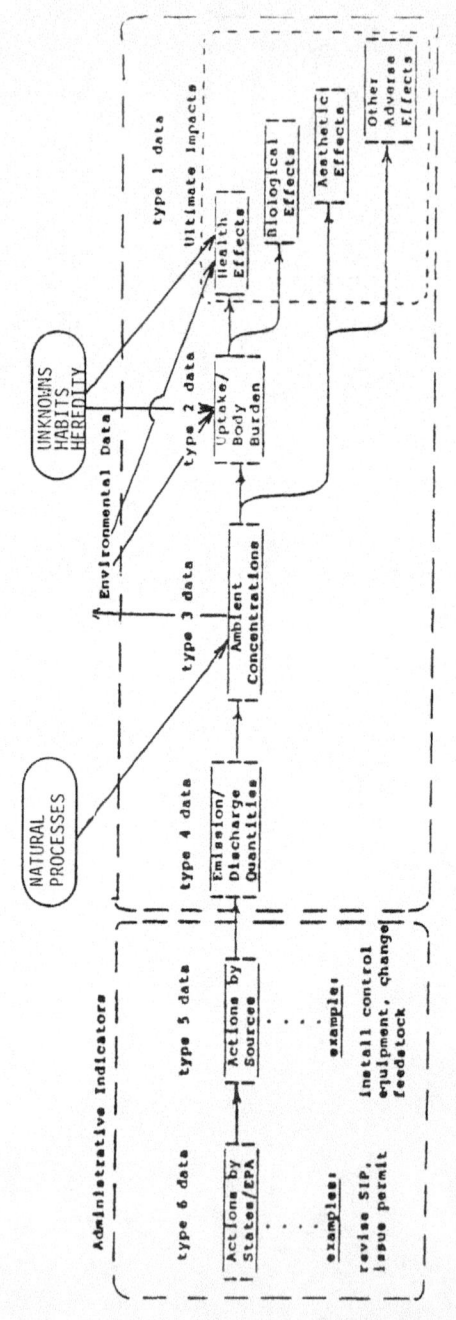

NOTE: Data to the right are colser to the "adverse ultimate impacts of pollution" that the States and EPA are charged with preventing or mitigating.

All else being equal, data further to the right are better indicators of environmental result than data further to the left.

MODEL DEVELOPMENT PROCESS

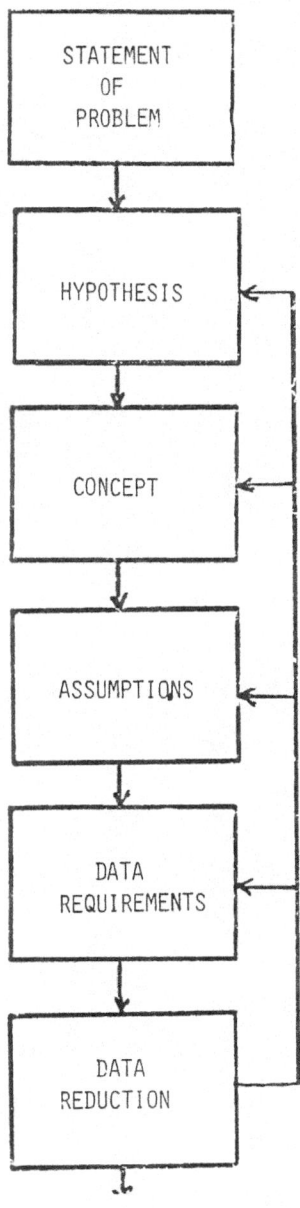

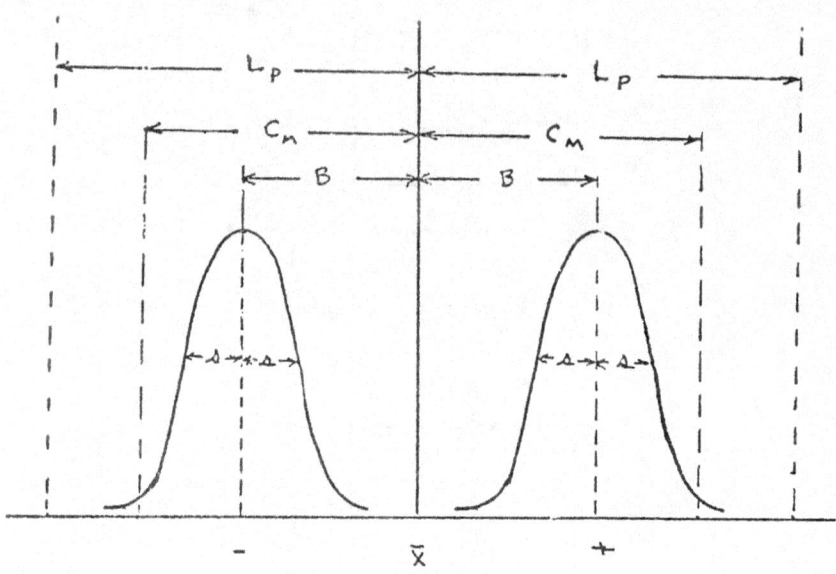

L_p = Tolerance Limit for Measured Property

B = Biases, Errors Inherent in Measurement

σ, S = Precision of Measurement

$\bar{\sigma}, \bar{S}$ = Precision of Mean of n Measurements

$$t\bar{S} = \frac{tS}{\sqrt{n}}$$

$$2\bar{\sigma} = \frac{2\sigma}{\sqrt{n}}$$

V

PLANNING AND SAMPLING

PLANNING

Who

    Analytical Chemist
    Client/Data User
    Statistician

What

    Sampling
    Measurement
    Data Format

Including

    Cost-Benefit Analysis
    Minimize Effort
    Maximize Information

Result

    Planned Experiment

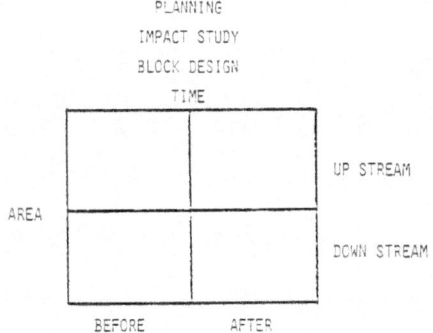
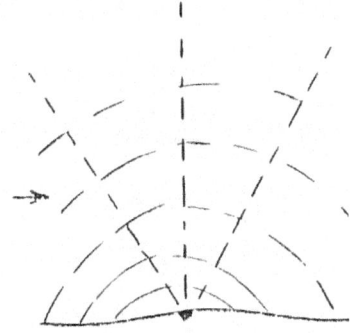

SAMPLING

MODEL

Objective
Kind
    Single Item
    Discrete Lot
    Defined "Bulk"
    Diffuse "Bulk"
Homogeneous
Heterogeneous
    General
    Stratified
    Localized
    "Needle-In-A-Haystack"
Composition
    Constant
    Variable (Time, Temperature, Space, Etc.)
        Predictable
        Unpredictable
Information Base
    Mean
    Extremes
    Distribution

PLAN

Type
    Random
    Systematic
    Representative
    Composite
    Subsample
Collection
    Method/Equipment
                        General
        Specific
    By Whom
        Analyst
        Collector
    Mode
        Spot
        Continuous
        Intermittent
        Automated
    Protocol
        Who, How, Where
        When, <u>What</u>

SAMPLING UNCERTAINTIES

Errors

    Random

        Sample Related
        Sampling Related

    Systematic

        Discriminatory
            Size, Mass, Etc.
        Inefficiency
        Collateral Measurements
            Time, Flow, Temperature, Pressure, Etc.

Deteriorations

    During Collection
    During Transit
    During Storage

Interactions

    Other Constituents
    Container, Transfer Lines

Stratification
Model Related

    Homogeneity
    Foreign Objects
    Moisture

SAMPLE TREATMENT

Homogenization
Reduction
Moisture/Drying
Preservation

TESTS FOR HOMOGENEITY

Measurement Plan

s_o, s_a, s_s

Statistical Tests

Within vs. Between Sample Variance

GUIDANCE IN SAMPLING

NUMBER OF SAMPLES (σ_A negligible)

$$n_s = \left(\frac{z\sigma}{E}\right)^2$$

or

$$n_s = \left(\frac{ts_s}{E}\right)^2$$

NUMBER OF ANALYSES PER SAMPLE

$$n_a = \left(\frac{z\,\sigma_a}{E_a}\right)^2 \qquad \sigma_s \text{ negligible}$$

If either n_s or n_a is too large

    In case of n_s,

        Take larger sample; composite; accept larger uncertainity

    In case of n_a

        Use more precise method; improve precision; accept larger uncertainty

WHEN VARIANCE OF SAMPLES AND MEASUREMENT BOTH CONTRIBUTE

$$E_{Total} = z \left(\frac{\sigma_s^2}{n_s} + \frac{\sigma_a^2}{n_s n_a}\right)^{1/2}$$

where n_a = no. of measurements per sample

IF SAMPLES COME FROM DIFFERENT "STRATA"

$$E_{Total} = z \left(\frac{\sigma_B^2}{n_B} + \frac{\sigma_s^2}{n_B n_s} + \frac{\sigma_a^2}{n_B n_s n_a} \right)^{1/2}$$

where n_B = number of strata
n_s = number of samples per strata

The various σ's may need to be evaluated by suitable experiments

Note that no unique solution is possible; several values of n_B, n_s, and n_a will satisfy above. Cost considerations may be required.

COST CONSIDERATIONS

n_s = total number of samples
n_a = number of measurements per smple
C_s = cost per sample of sampling
C_a = cost per measurement
C = total cost

$$C = n_s C_s + n_s \cdot n_a \cdot C_a$$

It can be shown that

$$n_a = \left(\frac{C_s \sigma_a^2}{C_a \sigma_s^2} \right)^{1/2}$$

and

$$N_s = \frac{z^2(\sigma_s^2 + \sigma_a^2/n_a)}{E^2}$$

If maximum allowable cost is C_m

n_a is obtained as above and substituted in the expression below to obtain n_s

$$n_s \leq \frac{C_m}{C_s + n_a C_a}$$

The values for n_s and n_a are substituted in earlier equations to calculate E, which may be higher than desired value, hence compromise may be required.

SIZE OF SAMPLE

Ingamel's Sampling Constant

$$K_s = W\left(\frac{\sigma}{\bar{x}}\right)^2 = WR^2$$

R = Relative standard deviation of sample composition

Sampling Materials in Discrete Units - ASTM E300

$$\sigma_x^2 = \frac{\sigma_b^2}{n_b} \cdot \frac{N - n_b}{N} + \frac{\sigma_w^2}{n_b n_w} + \frac{\sigma_t^2}{n_t}$$

σ_x = s.d., overall, of the mean
σ_b = s.d. of units in lot
σ_w = s.d. of samples within segment
σ_t = s.d. of measurement
N = number of units in lot
n_b = number of units selected (random)
n_w = number of within-unit samples
n_t = total number of measurements

DRYING

Moisture as a "Foreign" Object

    Free Water
    Occluded Water
    Bound Water

Options

    Ignore Moisture
    Dry Before Analysis
    Correct for Moisture

        Water by Drying
        Water by Analysis

REPORTING RESULTS

Sample Uncertainty <u>Must</u> be Stated

CHAIN OF CUSTODY

    Sample Safeguards
    Sample Subdivision
    Laboratory Records
    Sample Custodian

VI

METHODOLOGY

INTRODUCTION

GOAL OF MEASUREMENT

COST-EFFECTIVE USEFUL DATA

METHODOLOGY CHOSEN TO MINIMIZE ERROR

COST-EFFECTIVE APPROACH

MINIMIZE MEASUREMENTS
OPTIMIZE SAMPLES

INFORMATION REQUIREMENTS

HOW GOOD IS METHODOLOGY

ADEQUACY FOR GIVEN USE
MERITS OF COMPETITIVE METHODS

HOW GOOD IS THE DATA

HOW GOOD IS LABORATORY PERFORMANCE

NOMENCLATURE

Technique - Physical or chemical principle for characterizing materials of chemical systems.

Method - An assemblage of techniques; implies reduction to practice.

Procedure - Detailed instructions to permit replication of a method.

Protocol - Methodology specified in regulatory, authoritative, or contractual situations.

Absolute Method - Method in which characterization is based entirely on physical (absolute) defined standards.

Comparative Method - Method in which characerization is based on chemical standards (i.e., comparison with such standards).

Reference Method - A method of known and demonstrated accuracy.

Standard Method - A method of known and demonstrated precision issued by an organization generally recognized as competent to do so.

Standard Reference Method - A standard method of demonstrated accuracy.

Routine Method - Method used in routine measurement of a measurand. It must be qualified by other adjectives since no degree of reliability is implied.

Field Method - Method applicable to non-laboratory situations.

Trace Method - Method applicable to ppm range.

Ultra Trace Method - Method applicable below trace levels.

Macro Method - Method requiring more than milligram amounts of sample.

Micro Method - Method requiring milligram or smaller amounts of sample

CLASSES OF METHODS

Class	Precision/Accuracy	Nomenclature
A	<0.01%	Highest P/A
B	0.01 - 0.1%	High P/A
C	0.1 - 1%	Intermediate P/A
D	1 - 10%	Low P/A
E	10 - 35%	Semiquantitative
F	>35%	Qualitative

Note: For trace analysis, move classes down one step.

For Ultra-trace analysis, move classes down two steps.

The class must always be designated, since no other terminology defines or implies this characteristic.

NOMENCLATURE

Technique

    Pyhsical or chemical principle applied to analytical measurement.

Method

    Adaptation of a technique to a specific measurement problem.

Procedure

    Detailed steps for application of method.

        Precise Language

        Critical Steps Identified

        Calibration Detailed

        Principles Explained

        Detailed References

Protocol

    Mandated Methodology

Standard methods are really standard procedures

EXAMPLES OF ANALYTICAL NOMENCLATURE

Technique

    For example, Spectrophotometry

Method

    For example, Pararosaniline Method (West-Gaeke)

Procedure

    For example, ASTM-D2914

Protocol

    For example, EPA Method 625

HOW ACQUIRED

Technique

    Existing Technology

    Transfer of Technology

    New Technology

Method

    Revision

    Adaptation

    Novel

Procedure

    Modification

    Original

Application

    Utilization

    Standardization

MEANINGFUL MEASUREMENT METHODOLOGY

Essential Characteristics

    Sensitivity
    Specificity
    Precision
        ⎤
        ⎥ FIGURES OF MERIT
    Repeatability
    Reproducibility
    Range
        ⎦

Desirable Characteristics

    Large Dynamic Range
    Ease of Operation
    Speed
    Low Cost
    Portability
    Ruggedness

PERFORMANCE PARAMETERS

Type of Sample

Forms Determined

Range of Application

Limit of Detection

Biases

Interferences

Special Requirements

Calibration Limitations

Time Requirement

Precision

Other Limitations

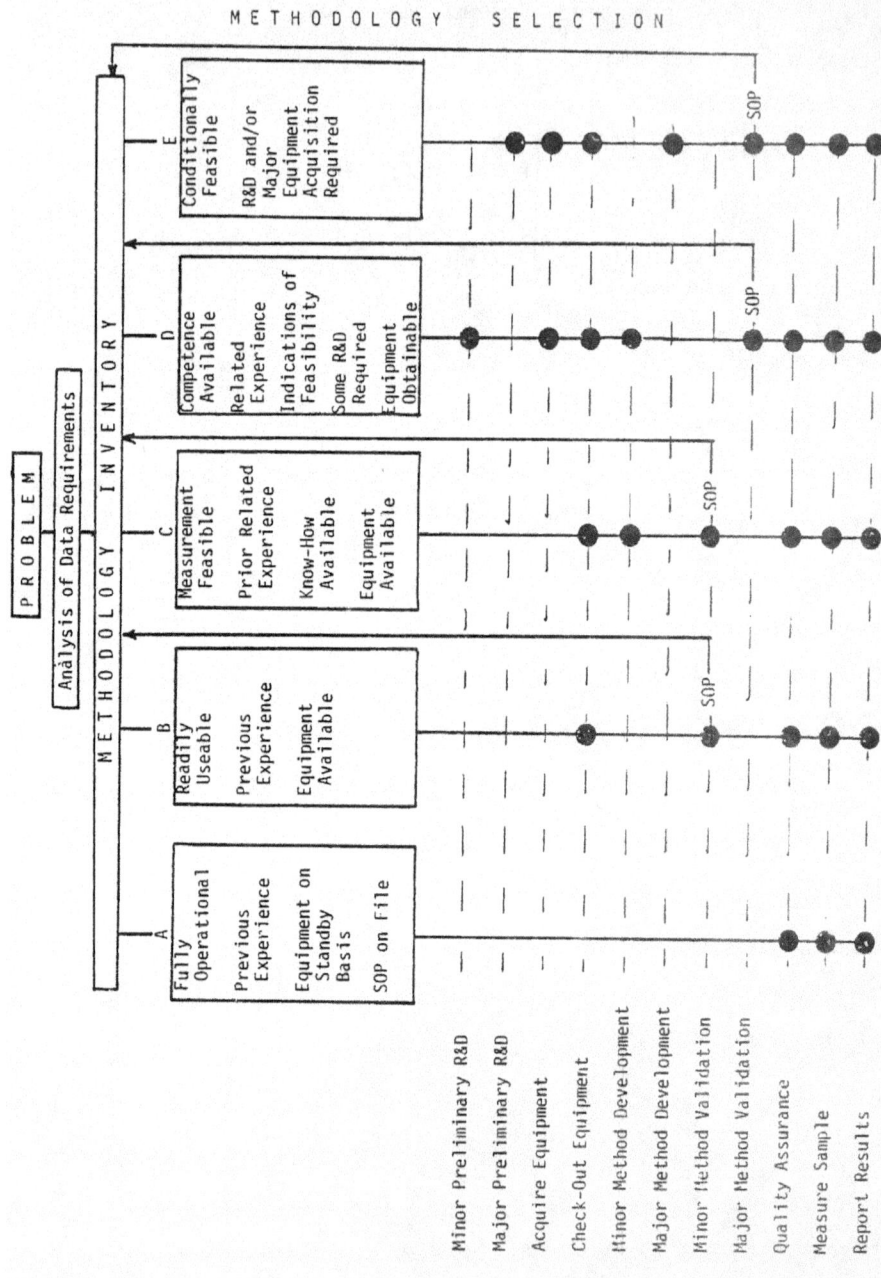

STANDARDS

Prepared by Organizations Recognized as Capable to Do So

Only as Reliable as the Organizations that Produce Them

Usually Result from Consensus of Opinion

Developed By

    Local - Special Organizations
    National Organizations
    International Organizations

Kinds of Standards

    Standard Methods
    Standard Practices
    Product Standards

All Developed as the Result of a Recognized Need

CHARACTERISTICS OF A RELIABLE STANDARD

Developed by a Reliable Organization
Developed by Representative Group
Developed by Consensus
Tested before Adoption

    Typical Process

Organization

    ↓↑

    Technical Committee

        ↓↑

        Subcommittee

            ↓↑

            Working Group

Advantages

    Based on Wide Experience, Pre-Tested, Well-Defined Means for Communication, can Produce "Standard Data."

Disadvantages

    May Hinder Innovation, Freeze Technology, May Induce False Confidence, May be Misused, May be Expensive -- Too Much Time to Develop.

PREREQUISITES FOR A TEST METHOD

Backed by research to define its characteristics
Validity supported by prior use
Ruggedness known or tested
Format chosen with

    End use/user in mind
    Degree of detail

        Self-contained
        For use with other information

            Question of adequacy of other information

REQUIREMENTS FOR COLLABORATIVE TEST MATERIAL

Matrix Match
Stability
Homogeneity
Method of Use

    Whole sample
    Sub-sample
    Spike

Shipability
Quantity

    For pre-testing
    For use in test
    For post-testing
    For post-use

Question of Blindness

    Blind
    Double-blind

COLLABORATIVE TEST OPERATIONS

Familiarization
Establishing statistical control*
Test measurements
Re-evaluation of Test Sample

---

*This is best shown by control charts based on pre-test measurements and maintained during the test.

RESULT OF COLLABORATIVE TEST

Test of a Method

    Pnecision

        Single operator/laboratory
        Between operator/laboratory

    Bias

        Level dependency

Proficiency Test

    Accuracy Evaluation

        Based on <u>true</u> value

    "Accuracy" Evalaution

        Based on consensus values
        Based on peer-group performance

CONCLUSIONS

Collaborative Test Results are Often

    Misunderstood
    Misinterpreted
    Poorly utilized

Precision/Accuracy of Method vs. Test of Method
Proficiency of Analyst/Laboratory vs. at Time of Test
Role of Standard/Validated Method
Importance of Statistical Control
Emphasis Should be on Identification and Minimization of Interlaboratory Bias

STANDARDIZATION OF A "METHOD"

TECHNIQUE → METHOD → PROCEDURE → PROTOCOL

1. Individual Laboratory

    Ruggedeness Test

       Procedure

          Single Operator Precision

2. Several Labs (3-9)

       Can Use Own Materials
       Evaluate - Comment

3. Comments → Revisions

       Devise Protocol for Testing

4. Collaborative Test on Same Material

       Unknown Composition
       Within Lab Precision
       Between Lab Precision
       Known Composition
       Bias
       Within Lab Precision
       Between Lab Precision

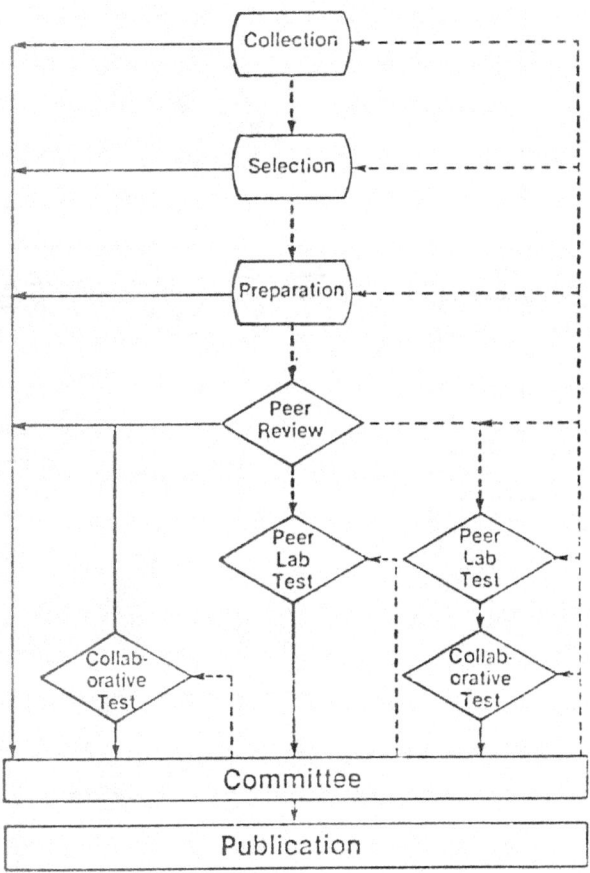

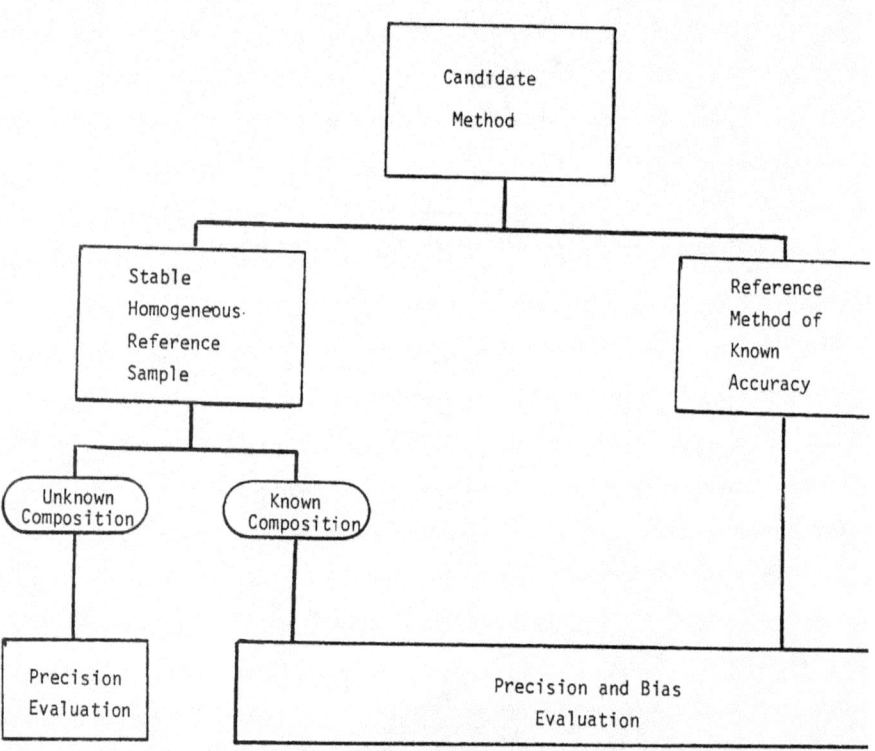

TESTING A STANDARD METHOD

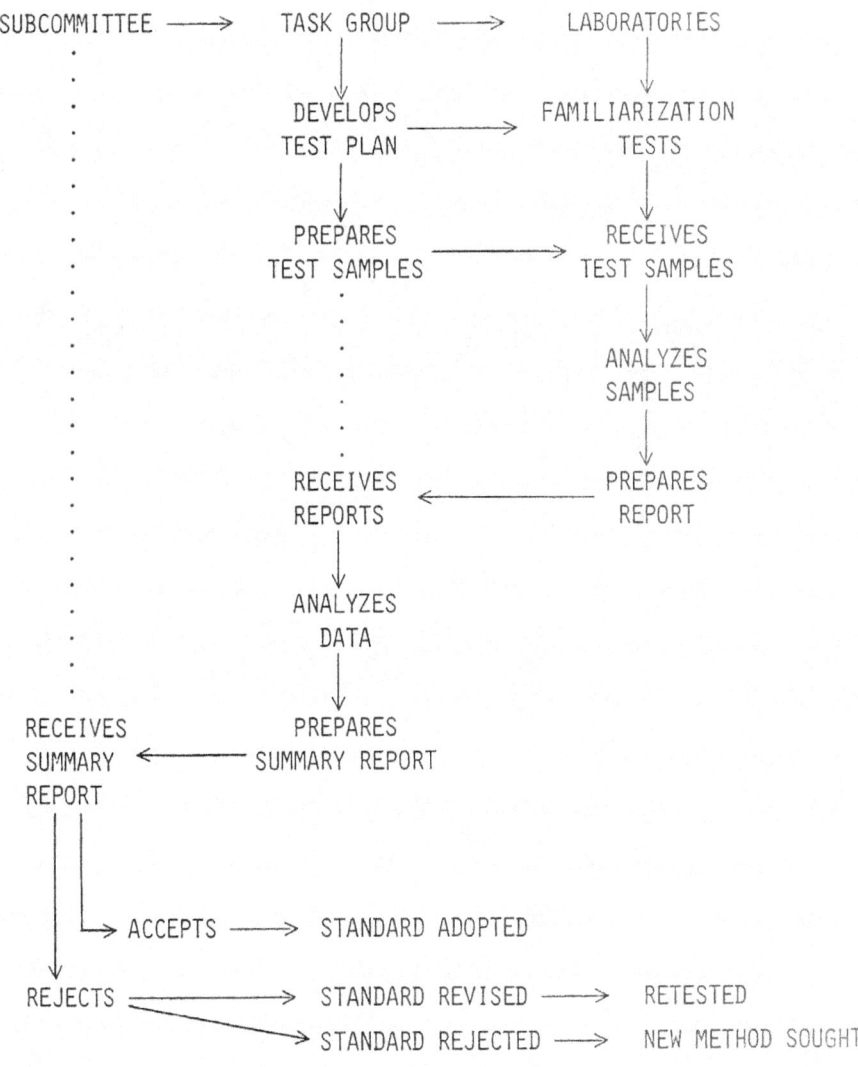

COLLABORATIVE TESTING

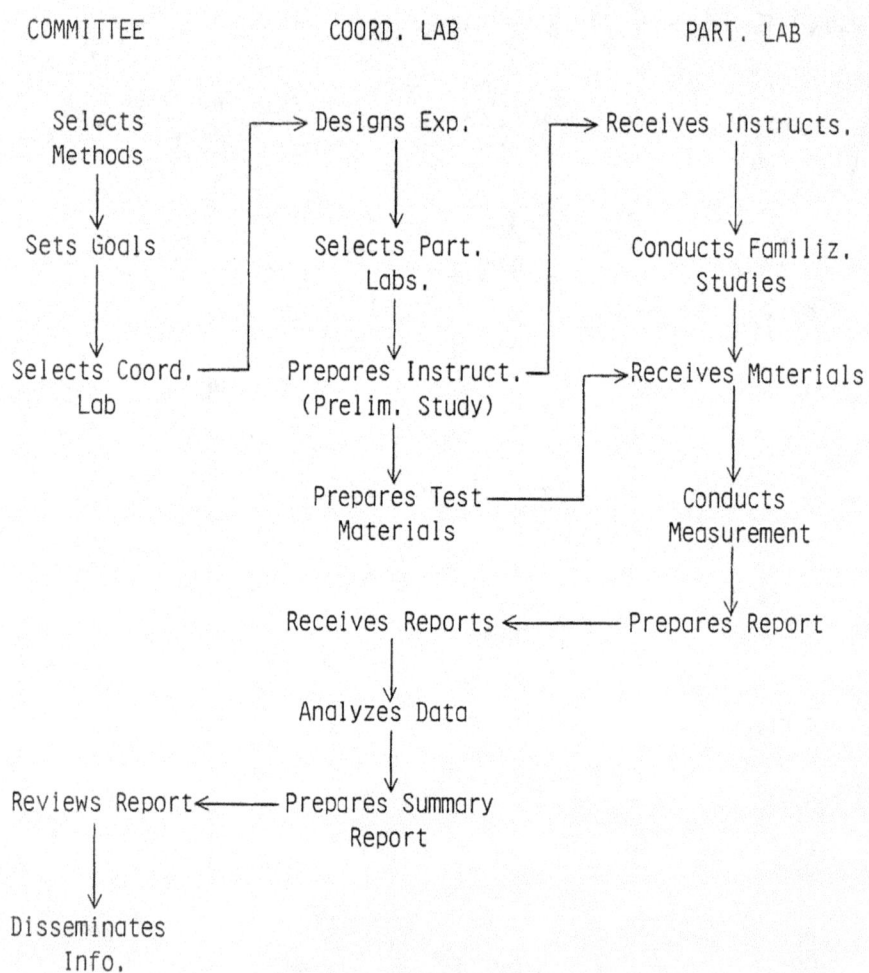

SIMPLIFYING COLLABORATIVE TESTING

System Concept of Measurement
- Identify critical steps
- Test performance of critical steps
- Improve methodology to reduce criticality

Matrix Concept of Measurement
- Test for matrix effects

Chemical Measurement Parameters

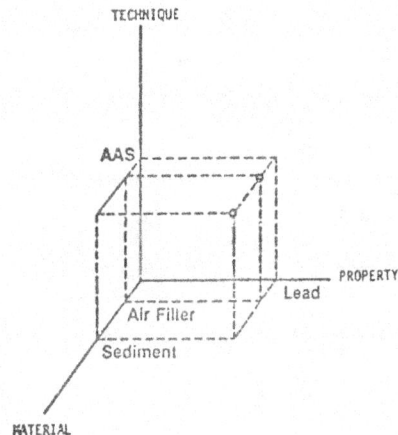

CHEMICAL MEASUREMENT PROCESS

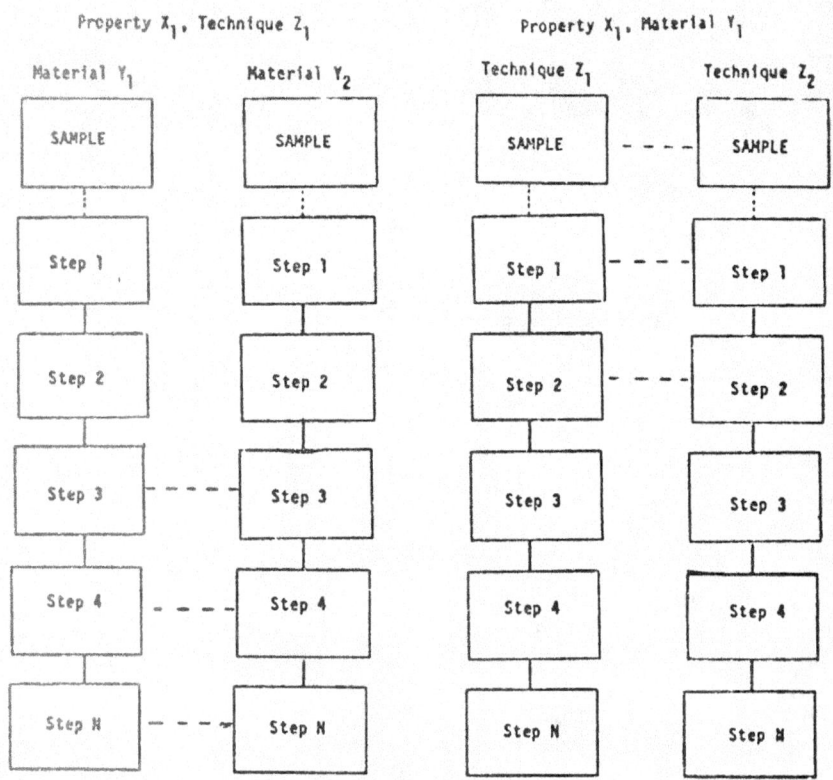

MATRIX MANAGEMENT CONCEPT

SIGNIFICANCE OF P AND A STATEMENTS OF STANDARD METHODS

Summarize the Collaborative Test

    Only as Good as the Test

    Influenced By:

        Design of Test Plan
        Kind and Number of Participants
        Prior Experience with Method
        Fidelity in Following Plan
        Fidelity in Following Procedure
        Quality of Test Materials

Give General Idea of Overall Performance Characteristics of Method

Indicate What Improvements can be Expected

Basis for Deciding Usefulness

Basis for Choice Between Competitive Methods

Basis for Comparing Your Performance with That of Others

Useful for Initial Setting of Control Limits

Useful for Initial Comparison of Isolated Results

Remember

    Standard Method Implies a Defined Situation for Its Use
    P&A Statement Applies Only to That Use

YOUDEN'S RUGGEDNESS TEST

Table 1. Eight combinations of seven factors used to test the ruggedness of an analytical procedure

Factor Value	Combination or Determination Number							
	1	2	3	4	5	6	7	8
A or a	A	A	A	A	a	a	a	a
B or b	B	B	b	b	B	B	b	b
C or c	C	c	C	c	C	c	C	c
D or d	D	D	d	d	d	d	D	D
E or e	E	e	E	e	e	E	e	E
F or f	F	f	f	F	F	f	f	F
G or g	G	g	g	G	g	G	G	g
Observed result	s	t	u	v	w	x	y	z

PROCEDURE

1. Chose 4 minus 4 combinations of $\underline{s}$ to $\underline{z}$ to get 4 caps minus 4 l.c. of desired letter.

   e.g. $\frac{s+t+u+v}{4} - \frac{w+x+y+z}{4} = $ A-a

2. Rank the seven differences to identify problems

3. Calculate s from the eight results, conventionally, or by $s = \sqrt{\frac{2}{7} \Sigma\, d^2}$ where d = A-a, e.g.

TABLE III.—SCHEDULE FOR TWELVE COMBINATIONS OF ANY NUMBER UP TO ELEVEN CONDITIONS.

1	2	3	4	5	6	7	8	9	10	11	12
A	A	a	A	A	a	a	a	a	A	a	a
B	b	B	B	B	b	b	b	B	b	b	B
C	C	C	c	c	c	c	B	c	c	C	c
D	D	D	d	d	d	D	c	d	d	d	D
E	e	e	e	E	E	e	d	E	E	E	E
F	f	f	f	f	P	f	e	f	P	F	F
G	g	g	G	P	f	G	f	G	C	g	g
H	h	h	h	g	g	H	G	h	P	h	h
I	i	i	i	h	H	I	H	i	h	i	i
J	j	j	J	J	J	J	I	J	i	J	J
K	k	k	K	K	K	k	J	k	J	K	k

W. J. Youden, in NBS Special Publication 300, H. Ku, Editor.

DETECTION LIMITS

IDL — Instrument Detection Limit

    Smallest "signal" instrument can reliably detect.

MDL — Method Detection Limit

    Smallest concentration/amount of analyte method can reliably detect, wherever located.

LOD — Limit of Detection

    Smallest concentration/amount of analyte that can be reliably reported as found/detected in a material/sample.

ACS GUIDELINES
LIMITS OF DETECTION/QUANTITATION

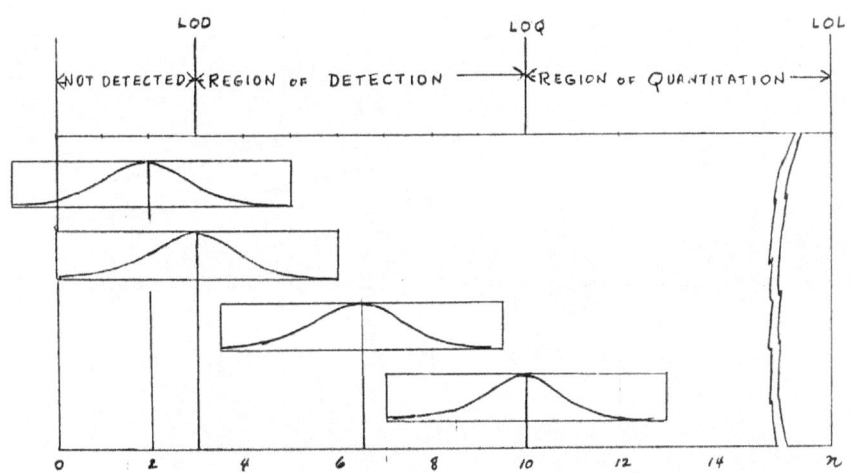

Measured Value in Units of Sigma (σ) with 99% Confidence limits
(LOL is Limit of Linearity)

VI-19

RELATIVE UNCERTAINTY

OF MEASURED VALUE

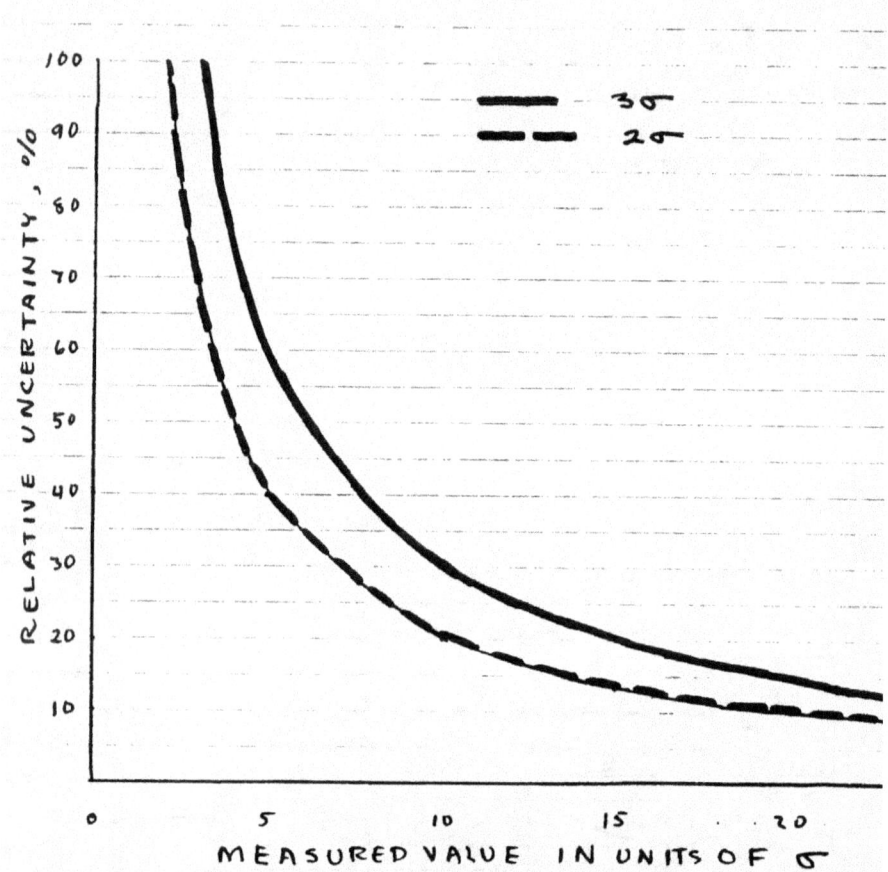

Measured Value in Units of σ

VI-20

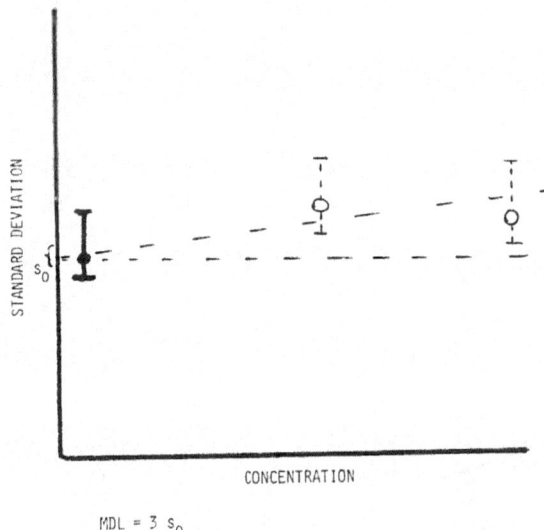

MDL = 3 s_0

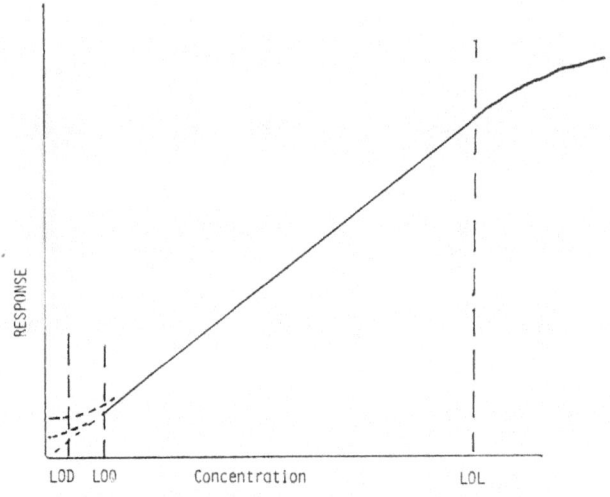

USEFUL RANGE FROM LOQ TO LOL

MAXIMUM HOLDING-TIME ESTIMATION

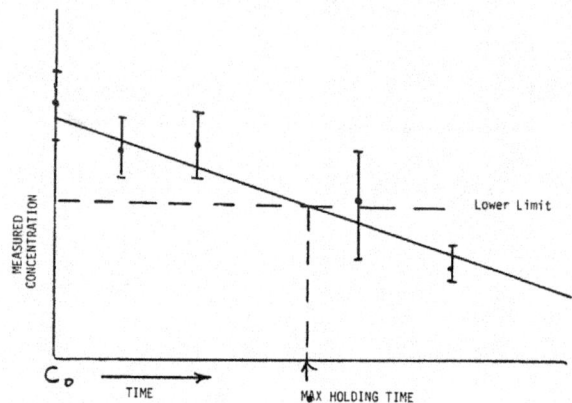

1. PLOT DATA AS FUNCTION OF TIME
2. DRAW BEST GRAPHICAL FIT (OR FIT BY REGRESSION)
3. CALCULATE $\bar{R} = \dfrac{d_1 + d_2 + \ldots d_n}{n}$ d = dif. of dupl.
4. CALCULATE $s = \bar{R}/d_2^* \approx 0.85\bar{R}$
5. CALCULATE LOWER LIMIT = $C_o - 3s \approx C_o - 2.55\bar{R}$
6. PLOT LOWER LIMIT AND DETERMINE TIME OF ITS INTERSECTION WITH DATA-FIT LINE
7. INTERSECTION TIME IS "MAXIMUM ACCEPTABLE HOLDING TIME" BY DEFINITION

VII

CALIBRATION

CALIBRATION

Physical

    Standards Used
    Frequency
    Slow Drift Followed by Abrupt Changes at Each Recalibration

Chemical

    Standards Used

        May Need to be Prepared by Experimenter

    Frequency
    Drift and Abrupt Changes as Above
    New Standards May Show Differences from Predecessors
    May Need Intercomparison

Calibration vs. Tolerance Testing

MOST IMPORTANT

ASPECT OF CALIBRATION

MATRIX MATCH

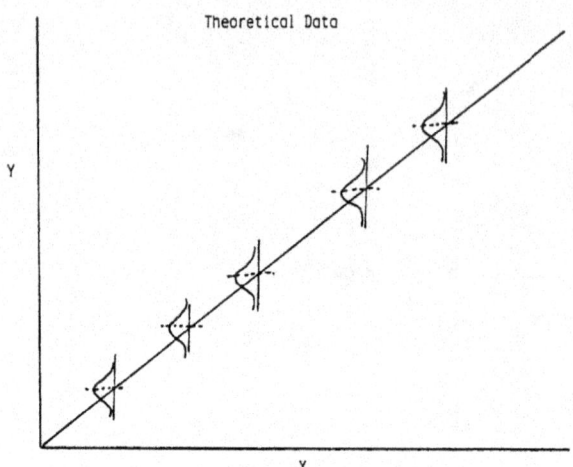

LINEAR FUNCTIONAL RELATIONSHIP, ONLY Y AFFECTED BY MEASUREMENT ERRORS

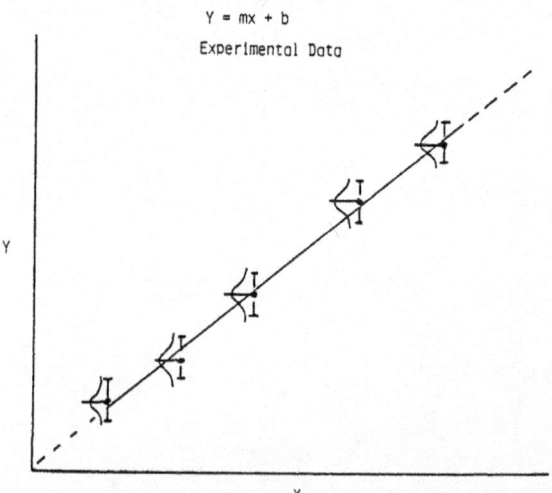

LINEAR FUNCTIONAL RELATIONSHIP, ONLY Y AFFECTED BY MEASUREMENT ERRORS

— Least Squares Regression Fit

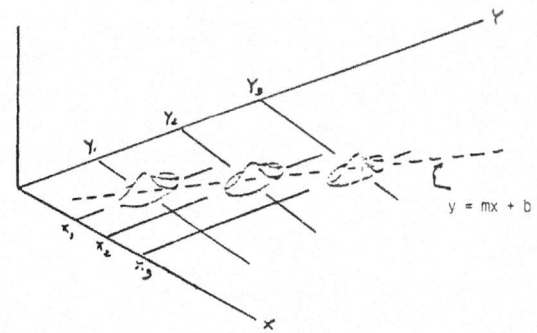

LINEAR FUNCTIONAL RELATIONSHIP, BOTH X AND Y AFFECTED
BY MEASUREMENT ERRORS

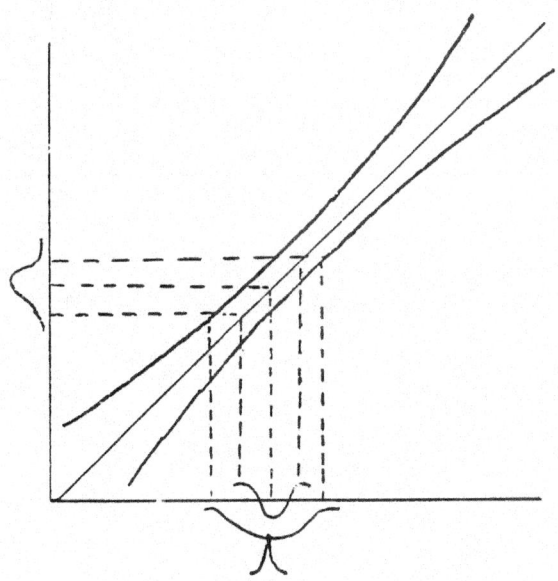

PROPAGATION OF CALIBRATION UNCERTAINTY

JOINT CONFIDENCE ELIPSE FOR SLOPE AND INTERCEPT

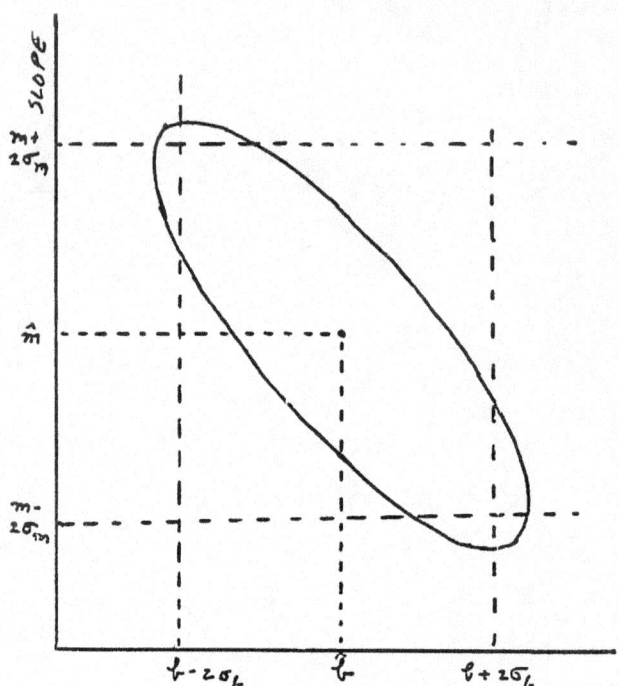

APPROACHES TO CALIBRATION

I. Matrix Independence

    A. Calibration by spiking/standard addition
    B. Calibration by representative matrix

II. Matrix Dependence

    A. Calibration by spiking/standard addition
    B. Calibration by matrix modification
    C. Calibration by matrix removal

        In Case II. Calibration must match methodology

MEASUREMENT IS A COMPARISON PROCESS

Measurement Consists of Comparison of Unknown with an Unknown

DIRECT COMPARISON

or

INDIRECT COMPARISON

e.g. Fixing the Value of a Scale (Calibration) for Direct Reading Instruments or Analytical Response Function for Others

Direct Comparison Can Be Considered As Continuous Calibration

Indirect Comparison is Usually Intermittent Calibration

CALIBRATION INTERVAL IMPORTANT

All Comparisons Require Analogy of Things Compared

Calibration Standards Must be Analogous to Samples Tested in Level and Matrix

SOURCES OF ERROR IN CALIBRATION

Standards Used

Linear Fit

Matrix Effects

    Can be Most Important Aspect of Calibration

    More Biased Measurements than Methods

    Biased Results from Unbiased Methods

Statistical Control Must be Demonstrated

VIII

QUALITY ASSURANCE

GENERAL

WHAT IS QUALITY?

Knowing Client's Needs

Designing to Meet Them

Faultless Implementation

Reliable Subcontractors

Punctual Delivery

Comprehensive/Understandable
   Reports/Interpretations

Back-up Service

> from
>
> Defect Detection
>
> to
>
> Defect Prevention

NEVER ENDING IMPROVEMENT OF QUALITY AND PRODUCTIVITY

OUTPUTS OF INCREASING QUALITY
WITH INCREASING PRODUCTIVITY

Dynamic vs Static

DETECTION TOLERATES AN ACCEPTABLE LEVEL OF DEFECT
PREVENTION AVOIDS DEFECTS

FUNDAMENTAL PHILOSOPHY

Current level of performance can be improved

Mental desire to do so

Room for improvement

vs.

Doing your best

The goals of yesterday are the commonplace occurrences of today and the outmoded practices of tomorrow

QUALITY ASSURANCE PROGRAM
WHY NEEDED

To discharge management's responsibility for quality of laboratory's ouputs

To satisfy analyst's concerns for quality work

To provide records and documentation for present and future use, and to protect all interests

QUALITY ASSURANCE

BASIC INGREDIENTS

QUALITY ASSESSMENT

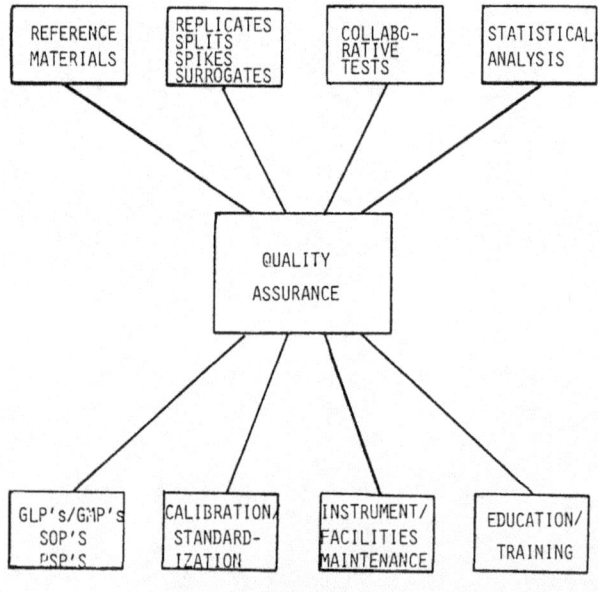

QUALITY CONTROL

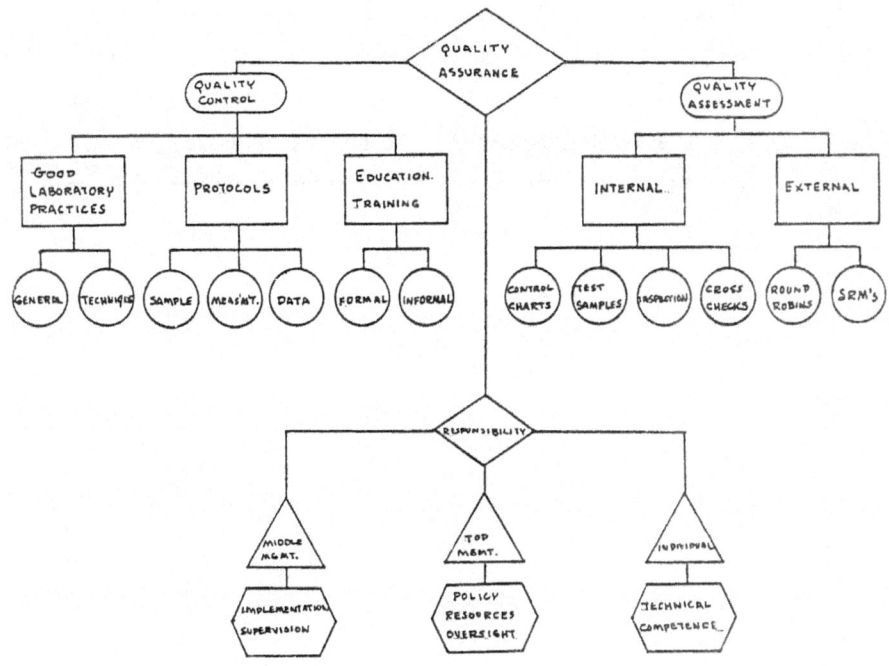

THE QUALITY ASSURANCE SYSTEM

"If (customer) expectations are not met—then all of the debates, the round-robin tests, the committee and task group work, and lofty statements of involvement with quality and commitment to excellence have not been productive."

IX

QUALITY CONTROL

"AN OBSTACLE THAT ENSURES DISAPPOINTMENT IS THE SUPPOSITION, ALL TOO PREVALENT, THAT QUALITY CONTROL IS SOMETHING YOU INSTALL. ...... ACTUALLY, QUALITY CONTROL TO BE SUCCESSFUL IN ANY COMPANY, MUST BE A LEARNING PROCESS ...... WITH ACCUMULATION OF KNOWLEDGE AND EXPERIENCE UNDER COMPETENT TUTELAGE."

>W. Edwards Deming
>"On Statistical Aids Toward
> Economic Production"
> Interfaces Vol. 5, No. 4, August
> 1975.

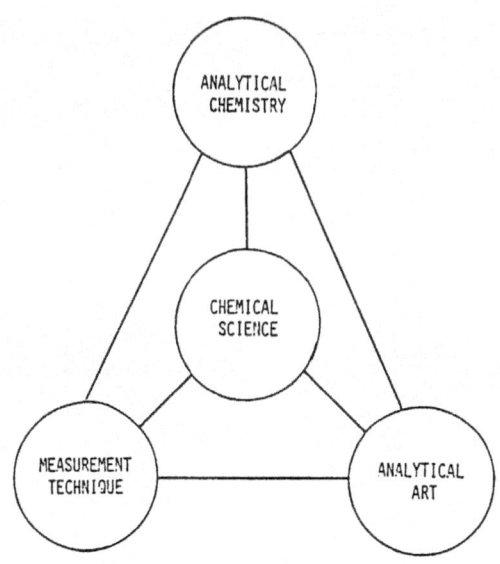

THE FOUNDATIONS FOR CHEMICAL ANALYSIS

A MEASUREMENT SYSTEM <u>MUST</u> HAVE ATTAINED A STATE OF STATISTICAL CONTROL

Randomness

Limiting Mean

Stable Variance

Fixed Distribution

UNTIL A MEASUREMENT SYSTEM HAS ATTAINED A STATE OF STATISTICAL CONTROL IT CANNOT BE CONSIDERED AS MEASURING ANYTHING.

QUALITY CONTROL

Basic Elements
    Good Laboratory Practices (GLP's)
    Good Measurement Practices (GMP's)
    Standard Operations Procedures (SOP's)
  Protocols for Specific Purposes (PSP's)
Education/Training
  Formal
    Courses
    Seminars
  Informal
    Discussions
    Readings
One-on-One
    Hands On

QUALITY CONTROL PROGRAM

Basic Elements
  Use of Qualified Personnel/Operators
  Use of Reliable Equipment
  Use of Appropriate Methodology
  Use of SOP's
  Strict Adherence to GLP's and GMP's
  Use of Control Charts
  Use of Appropriate Calibrations and Standards
  Close Supervision of All Operations by Management/Senior Personnel
  Protocols for All Critical Steps/Operations
  Protocols for Special Purposes

SOME FACTORS THAT INFLUENCE PRECISION

Operational Skill
Instrument Stability
Environmental Fluctuations
Reagent Control
Failure to Identify Critical Procedural Tolerances
Failure to Maintain Tolerances
Variable Recoveries
Variability in Control of Biases

SOME FACTORS THAT INFLUENCE BIAS

Interferences
Calibration
Inefficiencies
Losses
Contamination
Matrix Effects
Instrumental Shifts/Drifts
Resolution
Insensitivity
Operator-Related
Methodology Related
Application Related

GOOD LABORATORY PRACTICES

Address Such General Subjects

Laboratory Facilities
    Cleaning, Housekeeping
    Services-Temperature, Humidity
    Cleaning Glassware
Chemicals
   Grades, Storage, Handling, Disposal, Labeling, Shelf-Life, Water
Samples
    Custody, Documentation, Routing, Storage, Preparation, Retention
General Operations
    Weighing, pH
General Purpose Equipment
   Responsibilities, Use, Maintenance, Calibration
Data
   Reporting, Format, Release, Documentation
Statistical Procedures
Safety

GOOD MEASUREMENTS PRACTICES

For Each Measurement Technique, Address Such Subjects as
Maintenance of Equipment
    Records
Calibration
General Operations
Special Requirements Not Adequately Addressed by GLP's
Instruction Manuals
    Storage, Unkeep
Special Control Charts
Precautions

SOP'S
HOW BASIC OPERATIONS
are to be
DONE IN THE LABORATORY

SOP'S
for
SAMPLING
MEASUREMENT
CALIBRATION
DATA PROCESSING
STANDARD FORMAT ENCOURAGED
see XVIII - 9

EDUCATION AND TRAINING

Education - General
    Educational Deficiencies
    Advances in Technology
    New Technology
Training - Specific
    Initial and Continuing
    Job-Assignment Related
    Organizational Orientation
Safety
    Quality Assurance
    Change Related
    One-on-One
Serial Training Discouraged

QUALITY CONTROL BY INTERPRETATION

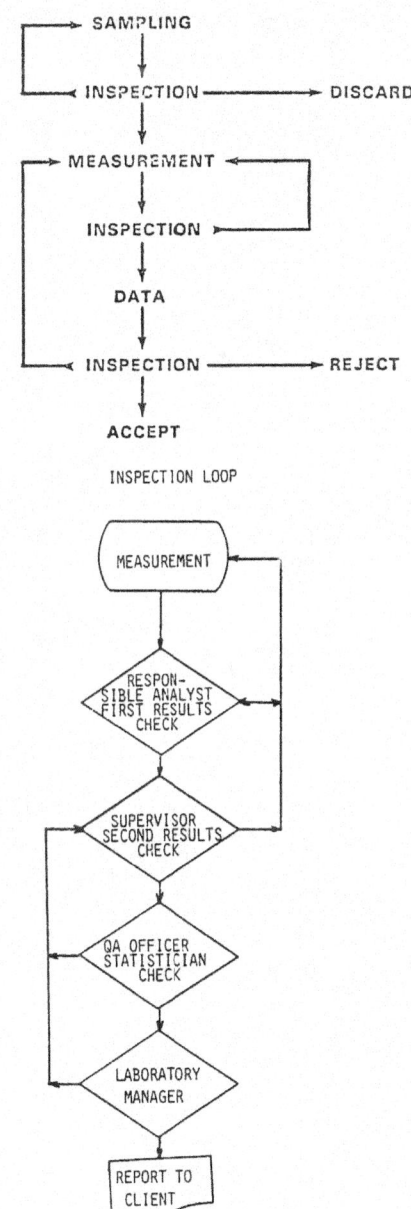

INSPECTION LOOP

YOU HAVE ALREADY FAILED IF YOU NEED A LOT OF INSPECTORS
YOU DON'T INSPECT QUALITY IN, YOU MUST BUILD IT IN

COMPUTERIZED INFORMATION MANAGEMENT

Measurement
- Sampling Numbering - Log-In
- Automatic Test Assignment/Work Orders
- Automatic Production of Labels
- Computer Filing of Methodology
- Easy Inspection of Previous Data
- Comparison of Results with Acceptable Runs
- Control Charts/Graphics
- Calibration Data - Checks
- Easy Inspection of Results
- Automated Report Generation
    - Reduction of Paperwork/Errors
- Filing/Archiving Data

Management
- Sample/Test Status Report
- Sample Location/Tracking
- Rapid Response of Inquiries
- Facilitation of Audits
- Records Management
- Statistics on Outputs
- Cost Accounting
- Chain of Custody
- Legal Defensibility

MAXIMS

The analyst should go over the data with the same care exercised in doing the analytical work.

o o o o o

The use of a computer does not relieve the chemist of the need to examine the data for suspect items.

o o o o o

Just glancing at data does not give one the familiarity that comes from working with the data.

o o o o o

Doing one's own computations whenever feasible is a great help to successful evaluation of analytical data.

GLP NO._____

Method for Cleaning Plastic Containers for Trace Element Samples*

Use: For containment and manipulation of aqueous solutions for inorganic trace analysis.

Materials: Containers should be constructed of conventional polyethylene (CPE), Teflon^R FEP or Teflon^R PFA. Teflon^R TFE is satisfactory but less desirable because of high surface porosity. Certain other materials are suitable but those cited here are the most commonly available through commercial sources. Containers should be free of visible occlusions within 1 mm of the surface. Closures should be fabricated of similar polymers (polypropylene, or Teflon^R) without cardboard liners or O ring gaskets.

Cleaning Procedure

Polyethylene: New -(polypropylene closure)

1) Using a clean room towel, wipe the outside of the container with solvent (hexane, an alcohol or ketone) and then distilled water to remove surface dirt. Use squeeze bottles to rinse the inside of the container with solvent and distilled water to remove as much surface dirt as possible.

2) Fill the container with a 1 + 1 dilution of reagent grade HCl. If facilities permit, submerging the container in a glass acid bath is preferable since it affords better cleaning of the closure threads and the outside of the bottle.

3) Allow to stand for one week at room temperature.

4) Empty container, rinse with distilled water and fill (as in (2)) with 1 + 1 reagent grade HNO_3.

5) Allow to stand for one week at room temperature.

6) Empty container, rinse several times with distilled water. Fill with highest quality distilled water and store until needed.

7) To use, empty, rinse with distilled water and dry under a clean (preferably Class 100) environment.

---

* Prepared by John R. Moody, National Bureau of Standards

Polyethylene - Used

1) The re-use of polyethylene bottles is not recommended unless they will be re-used for essentially identical samples. Any insoluble material in the bottle should preclude its re-use.

2) Repeat steps 1-3 plus 6-7. The leaching time in step 2 may be reduced to one day or less depending upon the degree of contamination of the bottle.

3) Note that polyethylene does not have a great tolerance for strong acid solutions and repeated cleanings may cause yellowing and embrittlement.

Fluoropolymers: New

1) Repeat steps 1-7.

2) Change temperature in steps 3 and 5 to 80 °C or more. Fluoropolymer surfaces when new have more surface impurities and require a more vigorous cleaning.

Fluoropolymers: Used

1) Unlike polyethylene and other plastics, the fluoropolymers are essentially inert and can be cleaned vigorously to remove contaminating traces of old samples.

2) Rinse or dissolve as appropriate any remaining sample material.

3) Repeat steps 1-7, changing the time and temperature in steps 3 and 5 to 8 hours or more at temperatures just below the boiling point.

1. Reference - J. R. Moody and R. M. Lindstrom, Anal. Chem. $\underline{49}$, 2264 (1977).

X

CONTROL CHARTS

CONTROL CHARTS

```
Purpose
    Criterion for Confirming Statistical Control or Its Lack
    Method of Identifying Assignable Causes
    Basis for Assigning Confidence limits for Data
Emphasis
    Order
        Sequence
        Time
    Grouping
        Place
        Source
        Test Conditions
            Measurement Variables
            Equipment
            Operators
```

TYPICAL CONTROL CHART

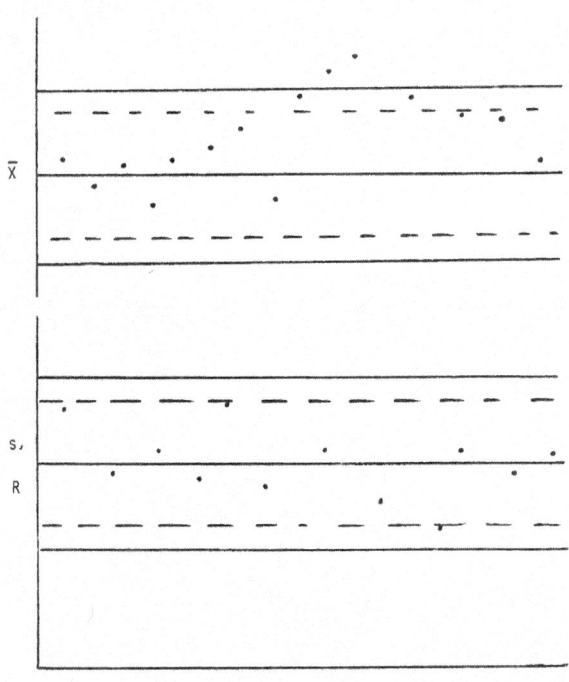

SEQUENCE/TIME

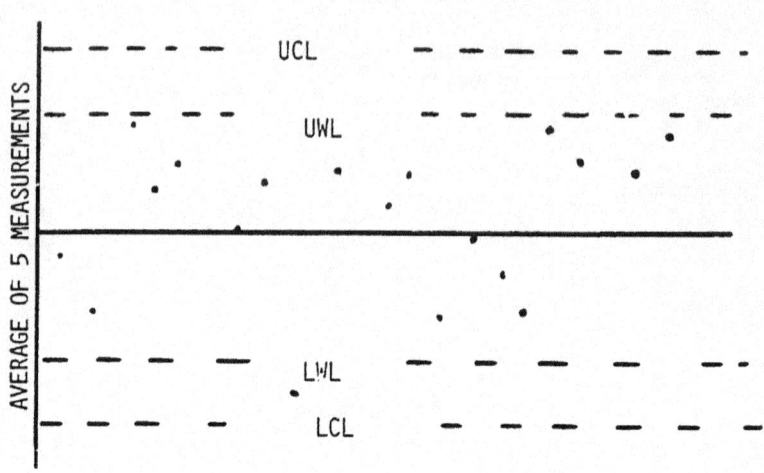

$\bar{X}$ CONTROL CHART

Advantages
    Useful For Non-Normal Distributions
        (means of non-normal distributions are essentially normally
           distributed)
    Takes Pressure Off Single Measurement
Disadvantages
    Increased Number of Measurements Needed
    Chance of Computational Error

CONTROL LIMITS

No Given Standard Available
    Based Entirely on Test Data
        Detect Inconsistencies
        Changes in Precision
           Changes in Accuracy
           Trends
           Cycles
        Detect Assignable Causes

Given Standard Available

    Based on Known $\bar{x}$, σ, R
    Standard Values Based on
    Representative Prior Data
        Desired Aimed-At Values
        Mandated Values
           Legal - Extrinsic
           Real - Intrinsic
Combination

CONTROL LIMITS

X-CHART

    Central Line $\bar{X}$

        WL $\pm 2\ S_B$

        CL $\pm 3\ S_B$

$\bar{X}$-CHART

    Central Line $\bar{\bar{X}}$, or known property

        WL $\pm 2\ S_B/\sqrt{n}$

        CL $\pm 3\ S_B/\sqrt{n}$

R-CHART - DUPLICATES

    Central Line $\bar{R}$

        UWL $2.512\ \bar{R}$

        UCL $3.267\ \bar{R}$

        LWL = LCL   0

SETTING CONTROL LIMITS
No Given Standard

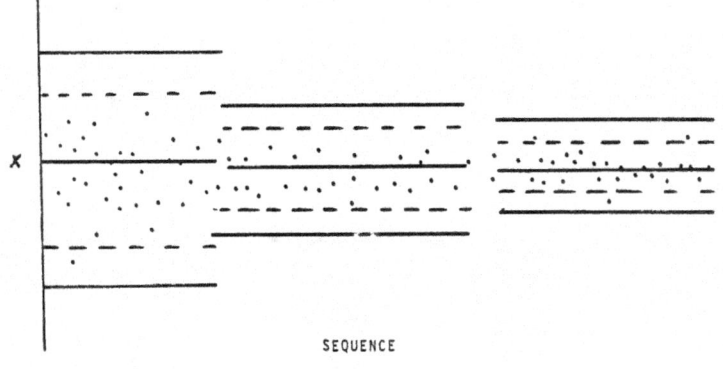

SEQUENCE

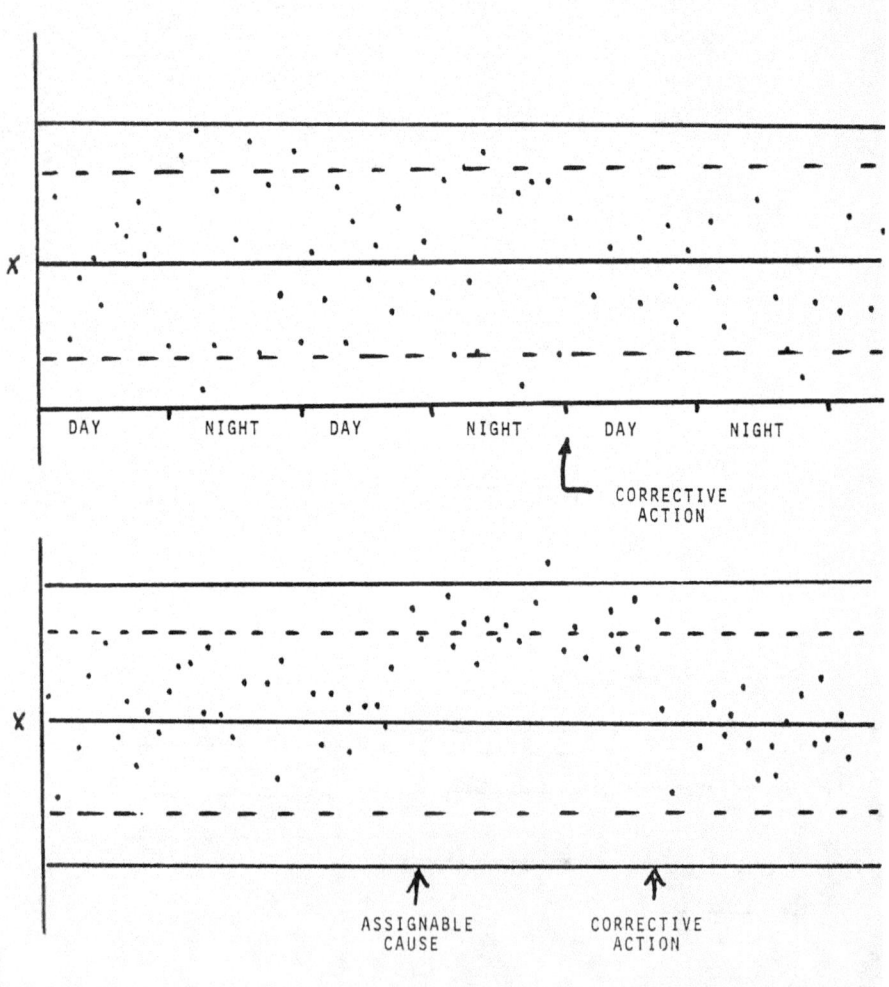

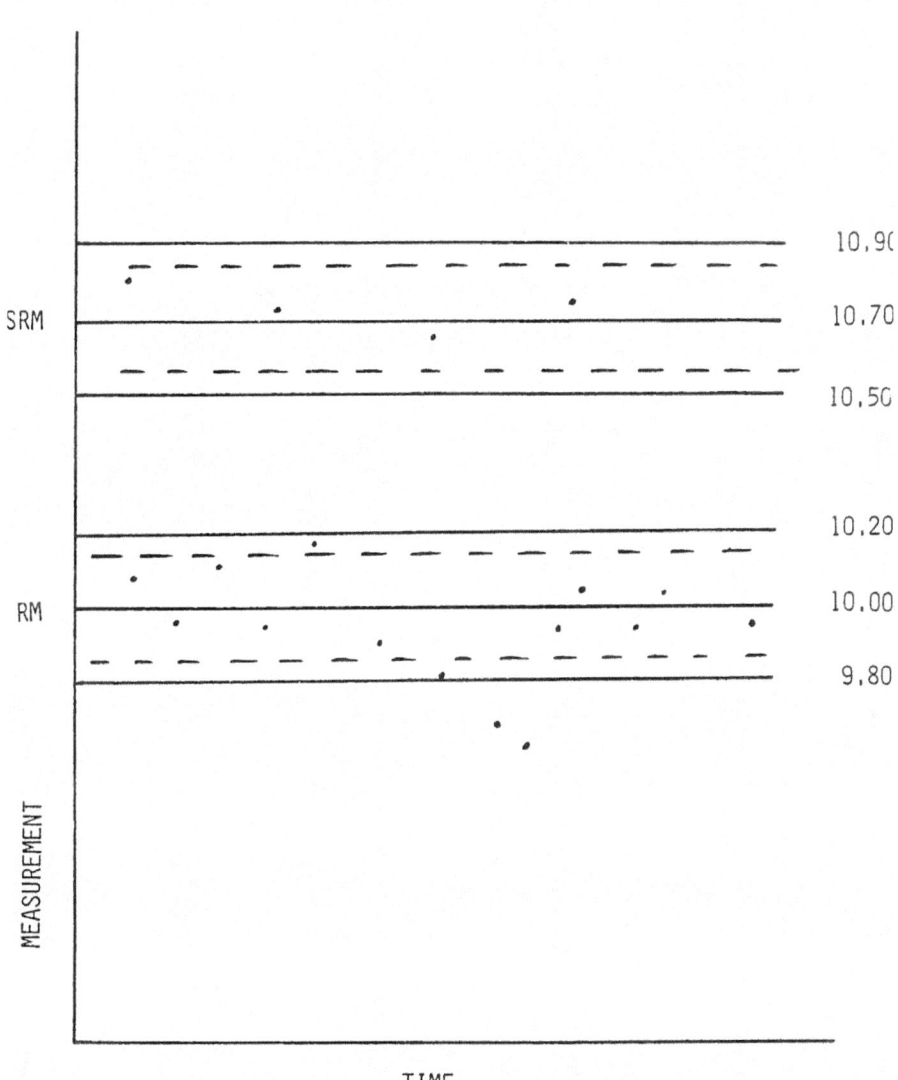

RANGE/DIFFERENCE CONTROL CHART

Idea

$$R \propto \text{Standard Deviation}$$

Approach
$$\bar{R} = (R_1 + R_2 + \ldots R_K)/K$$

For Duplicates

UCL = 3.267 $\bar{R}$ UWL = 2.512 $\bar{R}$
LCL = 0

Population of Concern
  All Measurements with <u>Similar</u> Character

  Relative Range, $\bar{R}/\bar{X}$, Permits Greater Inclusions

Advantages
  Applicable to Diverse Measurements Based on Performance of
    Actual Samples

Disadvantages
  Need for Replicate Measurements
  Possible Inclusion of Imcompatibles

DUPLICATE CONTROL CHART

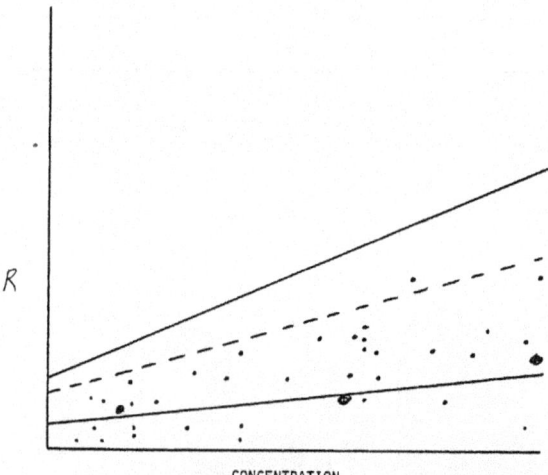

CONCENTRATION

FOR DUPLICATES
σ = 0.886 $\bar{R}$
$\bar{R}$ = 1.129 σ

UCL = 3.267 $\bar{R}$
UWL = 2.512 $\bar{R}$

RANGE RATIO CONTROL CHART

Range as a Function of Concentration

$$\bar{R} = fx$$
$$\bar{R} = k$$
$$\bar{R} = b + mx$$
$$\bar{R} = b + mx + cx^2 + dx^3 + \ldots\ldots$$

Relative Range, R_r

$$R_r = \frac{R_o}{\bar{R}} \;;\; R_o = \text{observed range}$$

For R

    Central Line = $\bar{R}$

    Control Limits = $D_3\bar{R}$ and $D_4\bar{R}$

For R_r

    Central Line = $\bar{R}/k$

    Control Limits = $D_3\bar{R}/k$ and $D_4\bar{R}/k$

    which leads to

    Central Line = $\bar{R}/\bar{R}$ = 1

    Control Limits = $D_3\bar{R}/\bar{R}$ and $D_4\,\bar{R}/\bar{R}$

                       = D_3 and D_4

Thus

    Central Line = 1

        UWL   = 2.512

        UCL   = 3.267

        LWL = LCL = 0

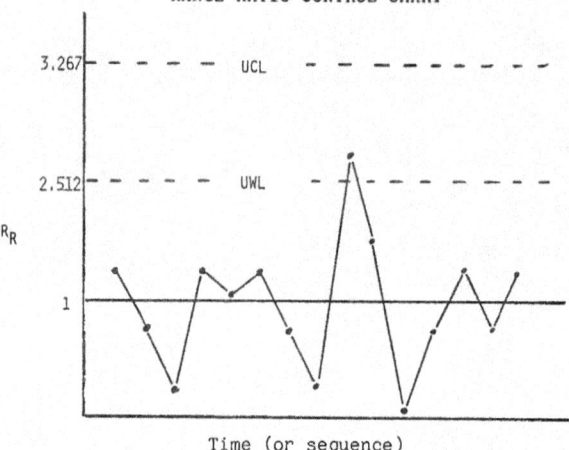

RANGE-RATIO CONTROL CHART

$R_r = \dfrac{R_0}{\bar{R} = fx}$

$\sigma \cong 0.886\ \bar{R} \cong 0.886\ fx$

$c.i._{.95} \cong 1.252\ fx$ (mean of duplicates)

$c.i._{.95} \cong 1.772\ fx$ (single measurement)

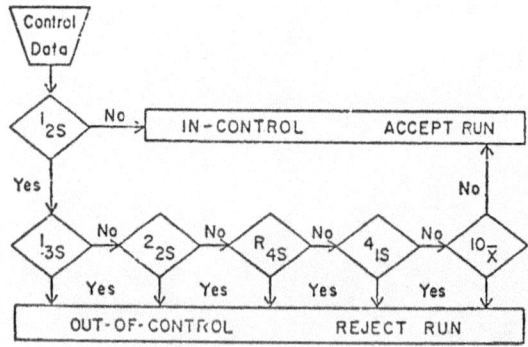

LOGIC DIAGRAM FOR APPLYING A SERIES OF DECISION CRITERIA (CONTROL RULES) IN THE MULTI-RULE SHEWHART PROCEDURE. (Westgard)

(See reference 25)

STRATEGY OF USE

Test Result Lies Outside Control Limits

$3 \sigma \equiv 3/1000$

Strategy of Runs

n	Expected
2	1 in 4
3	1 in 8
4	1 in 16
5	1 in 32
6	1 in 64
7	1 in 128
8	1 in 256
9	1 in 512
10	1 in 1024

Strategy of Seven

Seven consecutive values show rinsing tendency
Seven consecutive values show falling tendency
Seven consecutive values lie on one side of mean

Western Electric   Recommends Zones Concept

2 out of 3 in zone A
4 out of 5 in Zone B
1 out of 20 in Zone C
None out of Zone C
8 in a row in some pattern
Ascending
Descending
One side of Central Line

EXAMPLES OF CONTROL CHARTS

Selected SRM
IQA Sample
Typical Test "Solution"
Surrogates
Spikes
Duplicate Samples
Operator Charts
Instrument Operational Characteristics
Spectrophotometer Filter
Slope of Calibration Curve
Selected Calibration Point(s)
Extraction Recoveries
"Dummy" Test Object
Test Weight
Blank
Second Buffer

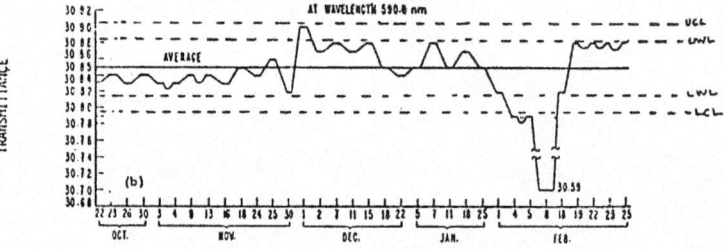

Control chart for a spectrophotometer showing the variation of transmittance as a function of time for a neutral flass filter at 635.0 nm (a) and 570.0 nm (b) and 24.0 °C.

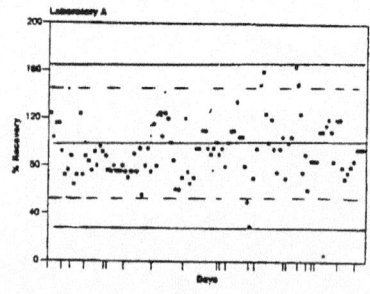
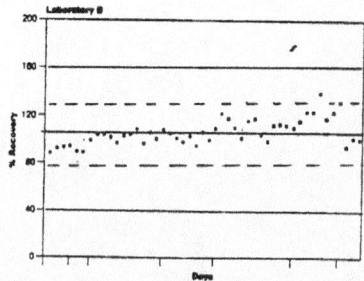

X-10

XI

QUALITY ASSESSMENT

QUALITY ASSESSMENT

Internal

    Internal Test Samples
    Control Charts
    Interchange of Operators
    Interchange of Equipment
    Repeat Measurements
    Independent Measurement

External

    Collaborative Tests
    Exchange of Samples
    Reference Samples
    SRM

INTERNAL TEST SAMPLES

Internal QA Samples (IQA)
Replicates
Splits
Spikes
Surrogates
Blind vs. Double Blinds

ALL BEST USED IN CONTROL CHART MODE

FREQUENCY OF QA SAMPLES
"Length of Run" Concept

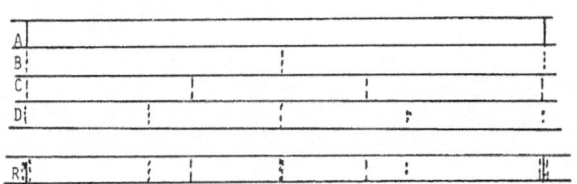

Sequence of Tests

Analyze Measurement Process for Variability of Sub-Steps
Estimate Variance of Sub-Steps
Use "Assignable Cause" Concept as Possible
Include at least two QA Samples, Preferably Three, in Each Run or Possible Run

QUALITY ASSESSMENT USING IQA SAMPLES

Daily/Event Schedule

<u>Calibration - Full Expected Range</u>

* IQA_0

   Test Samples - Group 1

* IQA_1

   Test Samples - Group 2

* IQA_2

   .
   .
   .
   .

   Test Samples - group N-1

* IQA_{N-1}

   Test Samples - Group N

* IQA_N

* Calibration - Midpoint

NOTES

    * - Decision Point
    1. Maintain Control Charts
       X - Control Chart, IQA
       R - Control Chart, ΔIQA
    2. System must be in Control at Decision Points
    3. At Least 2 Groups: Maximum of 10 Samples in Each Group
    IQA_i = occasion that a given IQA is measured.

QUALITY ASSESSMENT USING DUPLICATES/SPLITS
FREQUENCE SCHEDULE

Full Calibration

*   Calibration Check - Midpoint
    Sample 1

*   Sample 1 D/S
    Sample 2-9
    Sample 10

*   Sample 10 D/S
    Sample 11-19
    Sample 20

*   Sample 20 D/S
*   Calibration Check - Midpoint
*   Calibration Check - Midpoint/Duplicate

NOTES
    * = Decision Point
    1. Maintain R-Control Chart
       a. Duplicate Midrange Calibration
       b. Duplicate/Split Sample
    2. System Must be in Control at Decision Points
    3. If More than 20 Samples, Repeat Sequence
    4. If Less than 20 Samples, Divide into two groups and follow similar plan.

SPIKES

Internal Standards
    Carried through analysis
    Added before measurement
    Surrogates
    Analyte

Caution
    Standard Addition

Blank Correction
    See Appendix C

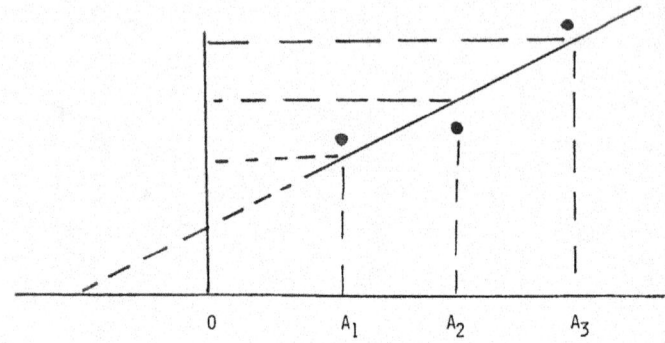

SPIKES

SLOPES SHOULD CHECK CALIBRATION
CURVE SLOPE

SPIKE/INTERNAL STANDARDS

Level Recommendations

Sample Concentration Expected
    < 10 MDL
    Spike at 20 x MDL increments

Sample Concentration Expected
    > 10 MDL

Routine Work
    Spike at 2X expected concentration

High Accuracy Work
    Iterative approach
    Run sample
    Spike at sample level, or
    Two spikes .90X and 1.10X sample level, using blank matrix

RECOVERIES

Add Known Amount and Measure

$$\% \text{ Recovery} = \frac{\text{Found}}{\text{Added}} \times 100$$

Use
- Efficiency Checks
  - Control chart use recommended
- Corrections
  - Coupled with control chart

Cautions
- Spike may not simulate sample
- Efficiency may be coupled to concentration level
- Variable recoveries indicative of trouble - lack of control
- Low recoveries signal losses
- High recoveries may indicate variable blanks, contamination
- Use of surrogates may not be definitive

REPLICATES

What is Replicated?

$$S^2_R = \Sigma\, s^2_{steps}$$

Samples

Entire Process

$$S^2_R = S^2_{sample} + s^2_{processing} + s^2_{measurement} + \cdots$$

Splits

$$S^2_R = \qquad\qquad s^2_{processing} + s^2_{measurement} + \cdots$$

Aliquots

$$S^2_R = \qquad\qquad\qquad\qquad s^2_{measurement} + \cdots$$

Should check with calibration replication
Suitable combination to isolate individual components
See "blank correction" for propagation of errors

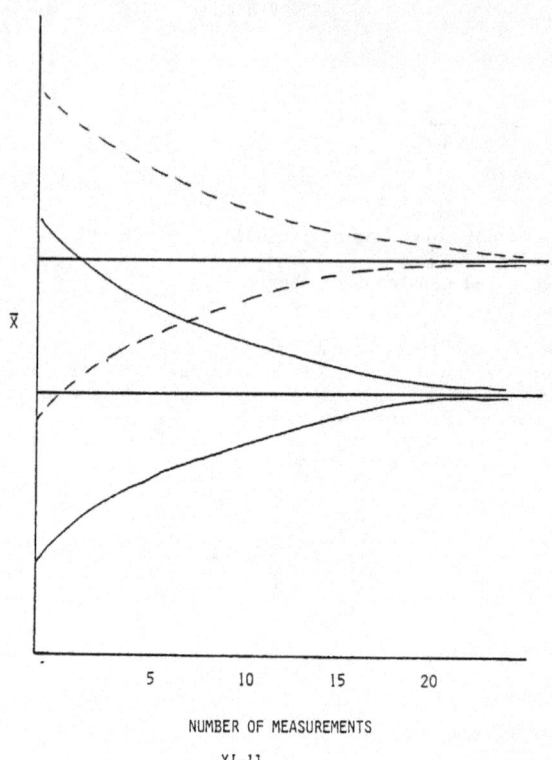

NUMBER OF MEASUREMENTS

XI-11

QUALITATIVE IDENTIFICATION

Inductive Reasoning
Known Selectivity
Knowledge of Absence of Possible Interferents
Experimental Demonstration
Confirmation by Independent Technique
Procedural Variations of Methodology
Combination

SYSTEM AUDITS

Internal and External

Spot Checks of Facilities

    Equipment
    Records
    Calibrations

Spot Checks of Selected Tests

    Some in detail
        for critical elements
    Some in general
        for entire process
    Identify defects
        classify as critical/non critical
    Take corrective action
        immediate remedies
        long-term actions
        modification of QA program

Internal Audits Minimize Surprises from External Audits

QUALITY AUDIT

Evaluation and Detection

    Standards
    Calibration
    Methodology
        Measurement-sampling-computations
    Reports
    Records

Assessments

    Discrepancies
        Significant-affect conclusions
        Minor-all others
    Categories
        Equipment
        Measurements
        Reports
        Records
    Feed Back
        Personnel Review
        Supervisory Review
        Management Review
    Corrective Actions

PERFORMANCE AUDITS
(Proficiency Tests)

Circulation of Test Samples
    Monitoring Networks

by   Regulatory Agencies
      Subscription Services

Information They Provide
    Typical Performance, or
    Best Performance

Either only if in a state of statistical control.

DEMONSTRATING ACCURACY

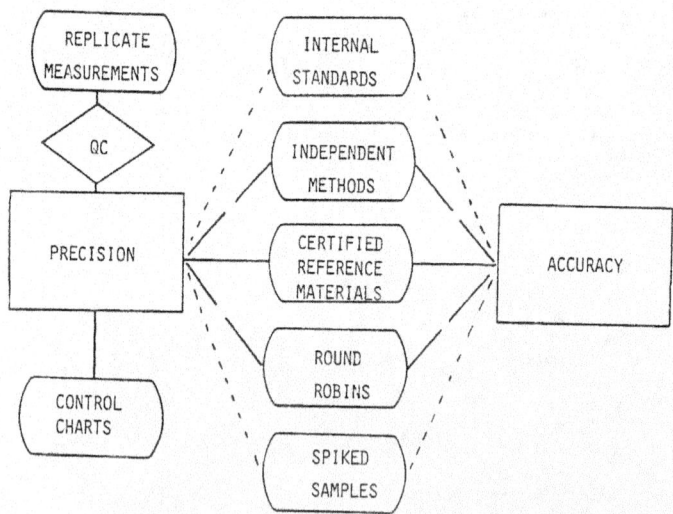

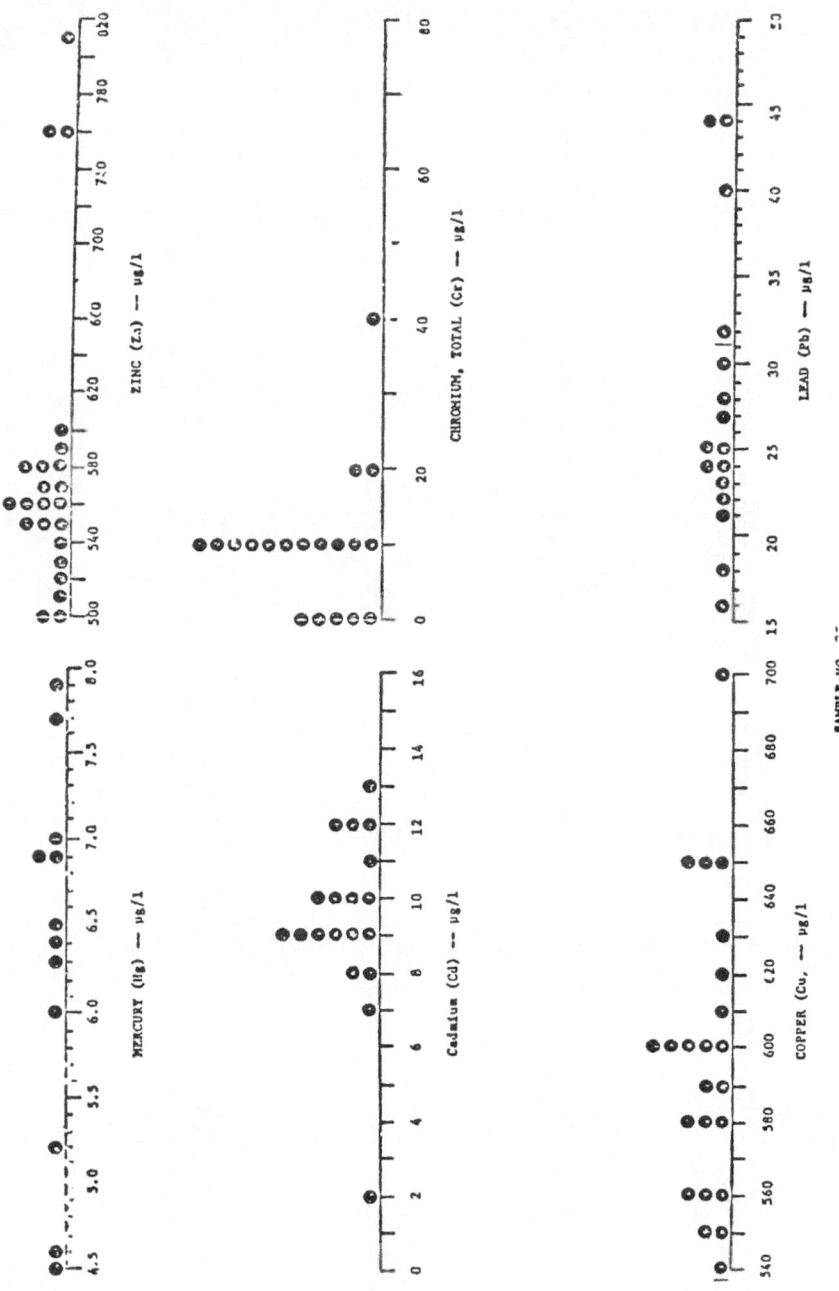

XI-9

ANALYTICAL RESULTS

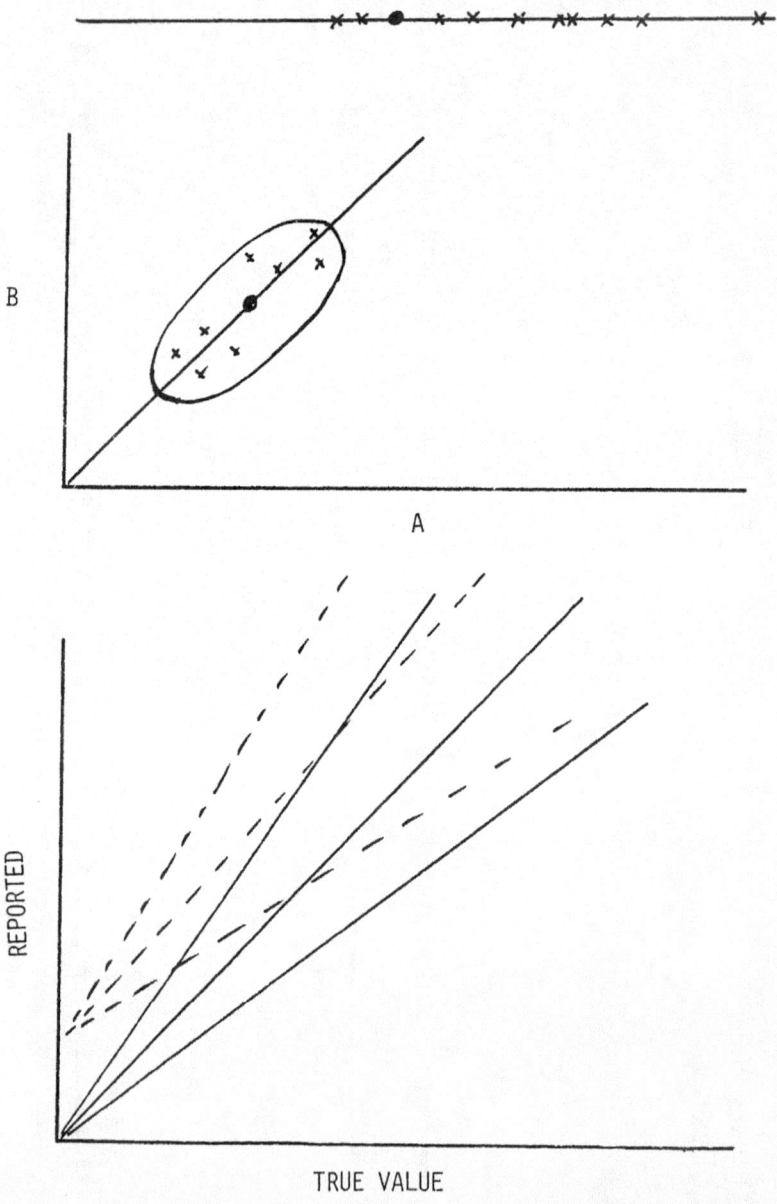

XI-10

TABLE I—Data and Calculations on Percent Insoluble Residue in Cement Reported by 29 Laboratories

Laboratory	Percent Residue		A − B	(A−B) − 0.095
	A	B		
1	0.31	0.22	0.09	−0.005
2	0.08	0.12	−0.04	−0.135
3	0.24	0.14	0.10	0.005
4	0.14	0.07	0.07	−0.025
5	0.32	0.37		
6	0.38	0.19	0.19	0.095
7	0.22	0.14	0.08	−0.015
8	0.46	0.23		
9	0.28	0.05	0.21	0.115
10	0.28	0.14	0.14	0.045
11	0.10	0.18	−0.08	−0.175
12	0.20	0.09	0.11	0.015
13	0.26	0.10	0.16	0.065
14	0.28	0.14	0.14	0.045
15	0.25	0.13	0.12	0.025
16	0.25	0.11	0.14	0.045
17	0.26	0.17	0.09	−0.005
18	0.26	0.18	0.08	−0.015
19	0.12	0.05	0.07	−0.025
20	0.29	0.14	0.15	0.055
21	0.22	0.11	0.11	0.015
22	0.13	0.10	0.03	−0.065
23	0.56	0.42		
24	0.30	0.30	0.00	−0.095
25	0.24	0.06	0.18	0.085
26	0.25	0.35		
27	0.24	0.09	0.15	0.055
28	0.28	0.23	0.05	−0.045
29	0.14	0.10	0.04	−0.055
Average	0.229	0.134	0.095	0.053

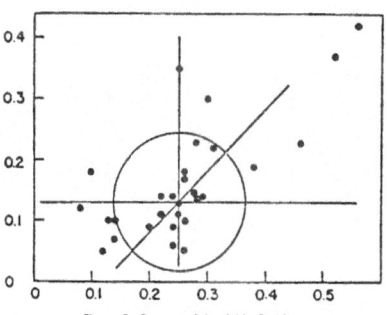

Figure 3—Percent of Insoluble Residue

TABLE II—Probability Table for Circular Normal Distribution

Percent of the Points Within Circle	Multiple b of the Standard Deviation
10	0.459
20	0.668
25	0.759
30	0.845
40	1.011
50	1.177
60	1.350
70	1.552
75	1.665
80	1.794
90	2.148
95	2.448
99	3.035

Note: Percent ≅ 100{1 − exp(−b²/2)}

Directions for Calculations
1. Tabulate Data Using Format of Table I.
2. Calculate A-B and The Average, $\overline{A-B}$.
3. Calculate $[(A-B) - (\overline{A-B})] = R$.
4. Calculate $\overline{R}$ = Average of Absolute Values of R.
5. Estimate σ, i.e., s by s = 0.886 $\overline{R}$.
6. Calculate 95% Confidence Circle Radius = 2.448s.

W. J. Youden, Ind. Qual. Control XV, No. 11 (1959)

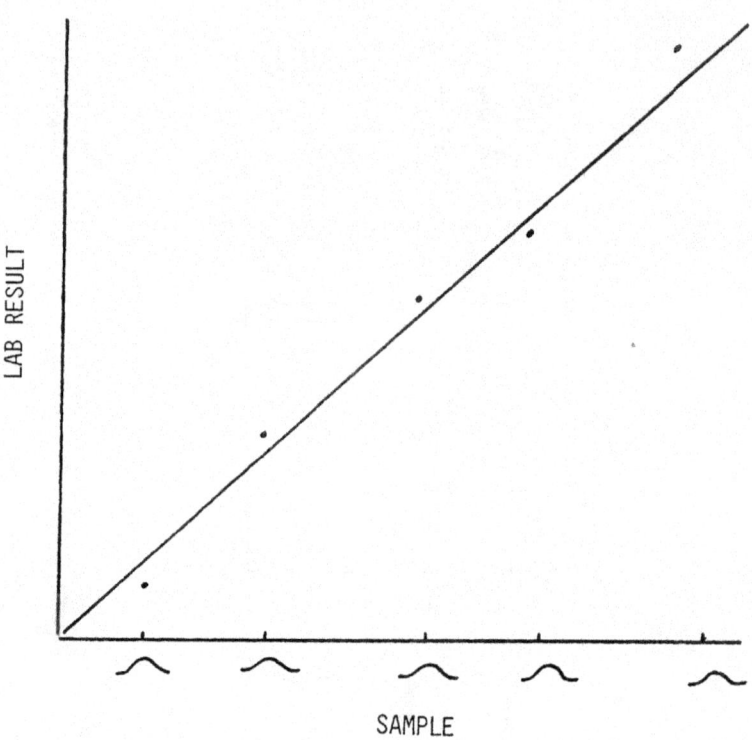

MULTISAMPLE

- Reduces Burden on Sample
- Identifies Range Dependencies
- Identifies Useful Range
- Detection Limit Information

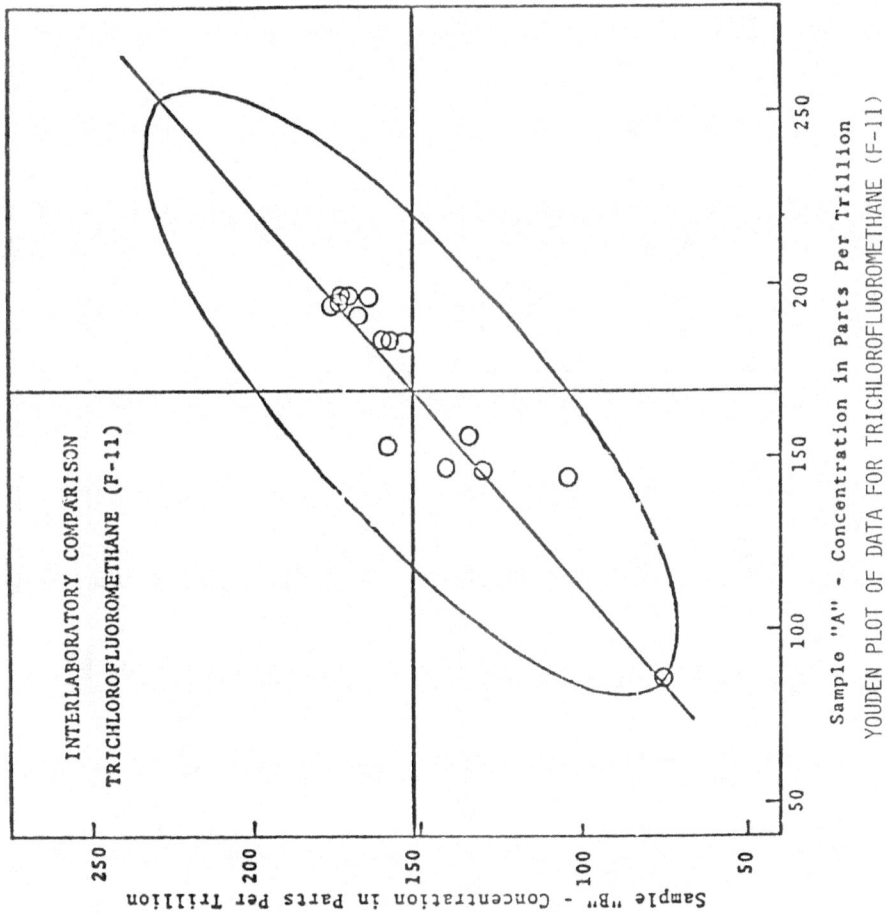

YOUDEN PLOT OF DATA FOR TRICHLOROFLUOROMETHANE (F-11)

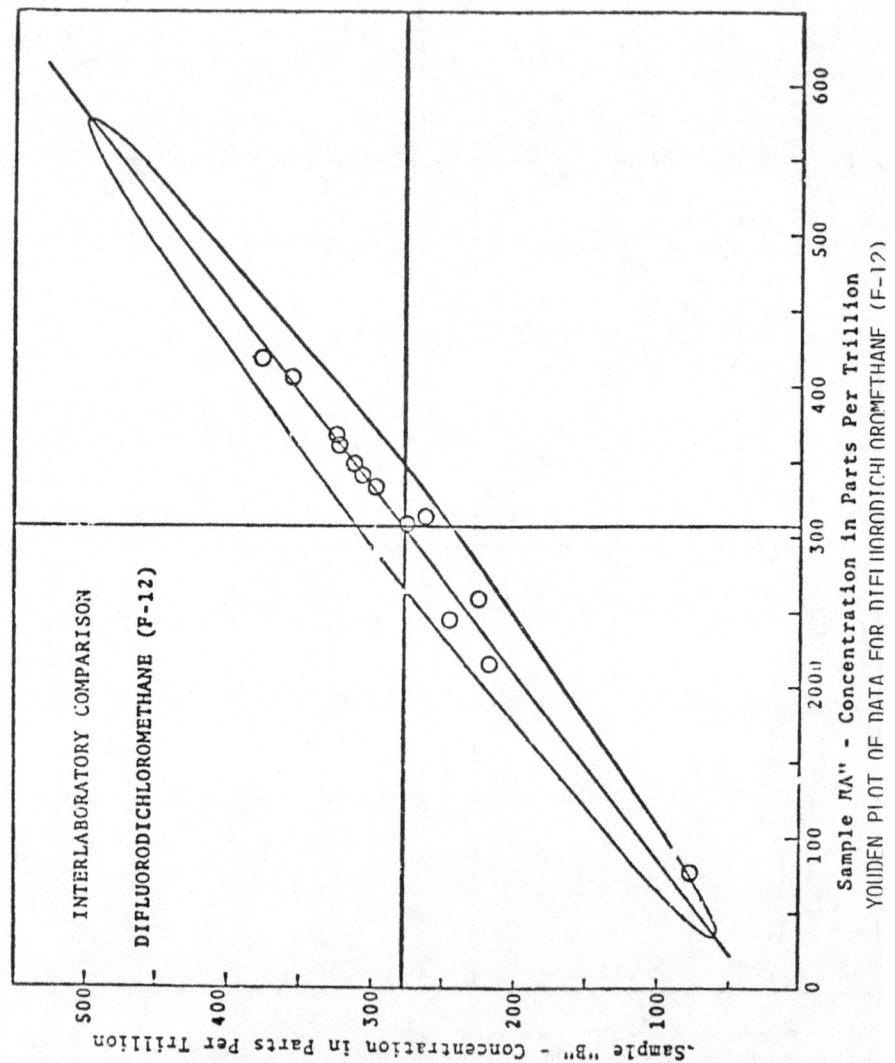

INTERLABORATORY COMPARISON
DIFLUORODICHLOROMETHANE (F-12)

YOUDEN PLOT OF DATA FOR DIFLUORODICHLOROMETHANE (F-12)

XII

CORRECTION OF ERRORS

AND/OR

IMPROVING PRECISION AND ACCURACY

MUST DISTINUGISH

between

THE TWO SOURCES OF ERROR
METHODOLOGY - INHERENT

and

APPLICATION - RELATED

TYPES OF CAUSES

CHANCES CAUSES
ASSIGNABLE CAUSES
COMMON CAUSES
SPECIAL CAUSES
SYSTEM CAUSES
PERFORMANCE CAUSES

THE DEMING DOCTRINE

Identify Defects
Tally Defects
Analyze Defects
Trace Defects to their Source
Make Corrections
Keep A Record of What Happens
Continue Until Defect is Eliminated

WHAT TO LOOK FOR

The Picket Fence Effect
    Using each preceding picket to measure the next, rather than a single picket as a model.

Think Small
    Don't overlook the small errors. They happen more frequently than large ones.

PARETO DIAGRAM

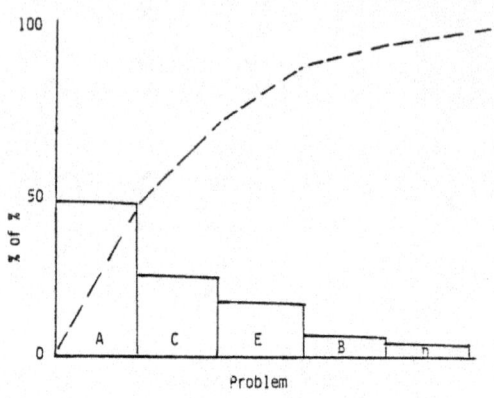

Problem	Occurrence	%Occurrance	% of %
A	198	9.1	47.6
B	26	1.2	6.0
C	103	4.8	24.7
D	18	.8	4.3
E	72	3.3	17.3
Total	416	19.2	99.9

2166 objects inspected

ISHIKAWA'S CAUSE-EFFECT DIAGRAM

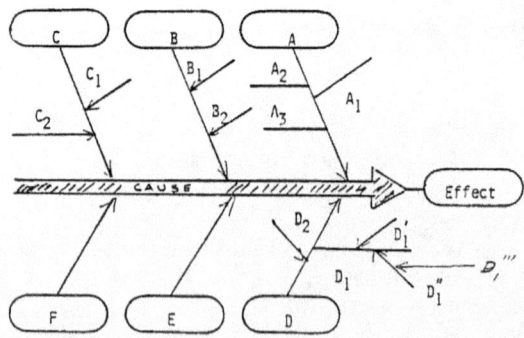

DEFICIENCY CORRECTION

When
    A. In control, Results Questioned or Questionable
    B. Out of Control

Check List

	A	B
Changes	✓	✓
Sampling	✓	
Sample Handling	✓	
Analytical Procedure		
Calculations	✓	✓
Data	✓	✓
Reagents		✓
Equipment		✓
Calibration		✓
Maintenance		✓
Methodology		✓
Blunder	✓	

ASSIGNABLE CAUSES

Based on Control Chart

Nature
    Bias
    Imprecision
    Uncertain

Occurrence
    Permanent
    Transitory

Attack Imprecision First

ERROR ANALYSIS

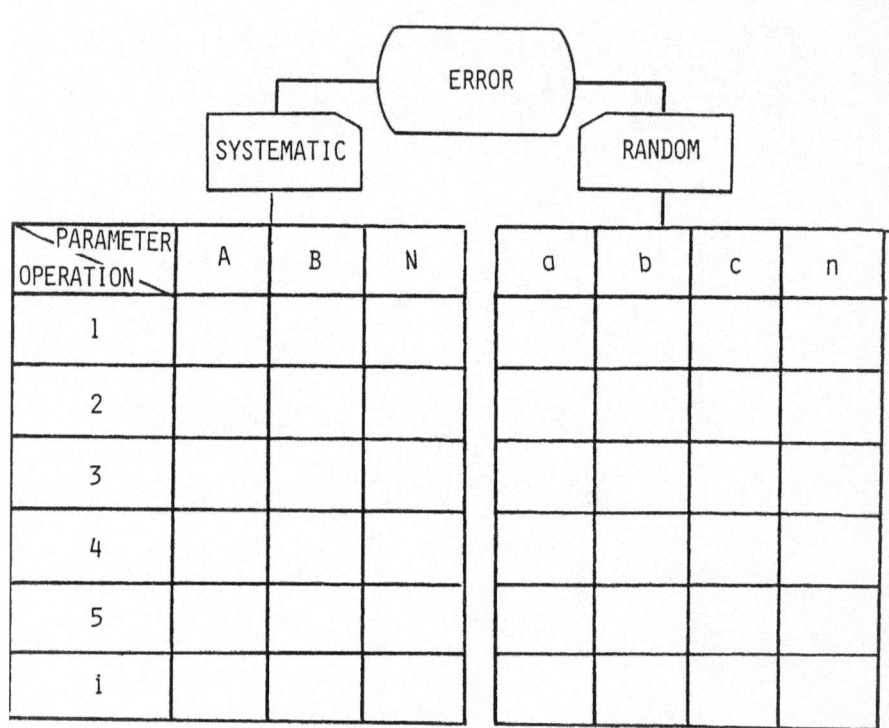

ANALYSIS OF SOURCES OF ERROR IN A TYPICAL CHEMICAL MEASUREMENT

PARAMETER / OPERATION	VOLUME	TEMPERATURE	TIME	MASS	CHEMICAL PURITY	LENGTH	MATRIX EFFECTS	MECHANICAL LOSS
SAMPLING	X		X		X			
EXTRACTION	X							X
DRYING		X						X
CONCENTRATION		X						X
CHEMICAL TREATMENT	X			X	X			X
MEASUREMENT	X	X	X			X	X	
CALIBRATION	X	X	X	X	X	X		
CONFIRMATION					X		X	
CALCULATION								
REPORTING								

PRELIMINARIES

Familiarization
    Dry Runs
        Use of Reference Materials

Analyst/Operator Check-Out
    General Proficiency
    Special Proficiency
        Training
            No Serial Training
        Pilot Runs

Performance Checks
    Availability of All Needs
    Stabilize Equipment
    Tune Equipment
        Optimize
    Verify System Performance
        Use of Test Sample
    All Systems Go!

"No quality assurance program, whether it be voluntary or imposed, can correct frequent mistakes and unreliable performance introduced by insufficient training, inadequate laboratory environment, and poor administrative practices."

                                          William Horwitz
                                          Quality Assurance Practices for
                                            Health Laboratories, p. 547
                                          APHA, 1978

KINDS OF PEOPLE

No concern for quality. Just want to get the job done without too many complaints.

Those in Between

Overconcerned about quality. Care so much the job never gets done.

Always trying to do better. "Just a few more measurements."

Quality Assurance Practices Help All Of These

LABORATORY PERSONNEL

The Most Important Aspect of Quality Control*

Laboratory Staff

    All Those Who Can Influence the Correctness of the Information

        Management
        Supervisors
        Analysts
        Technicians
        Support

Quality of Output Highly Dependent On

    Motivation
    Performance

Number of Mistakes/Analyst/Year

    34 Class A made 17% of Total
    19 Class B made 58% of Total
     3 Class C made 25% of Total

Training is Essential

    Technical Training to Provide

        Competence/Skills

    Supplemental Training/Introduction

        Personnel Must Have

            Appreciation of Interested Client
            Appreciation of Critical Aspects of the Work
            Appreciation of Personal Responsibilities

---

* J. A. Lott, Med. Instrumentation 8: 22-25; 1974.

WHAT INFLUENCES THE ANALYST

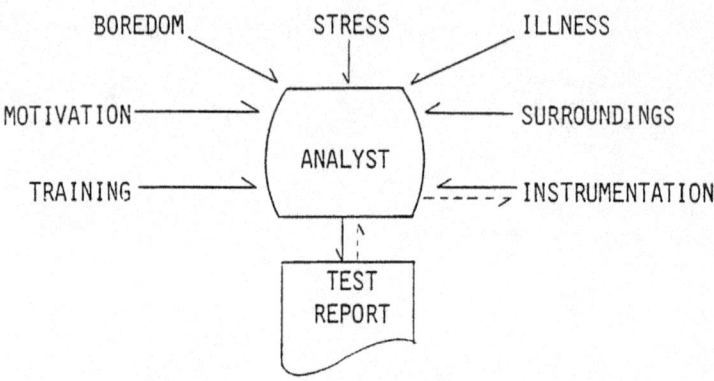

Training - To Provide Competence/Skills

Motivation - Influences How Carefully One Works

Boredom - Challenge Must Be Present

Stress - Work-Pressure Contributions

        Equipment Contributions

            Delicate, Difficult to Adjust, Drift, Tempermental, Unsuited

Illness - Physical/Mental

Surroundings - Space/Clutter/Cleanliness

Instrument - Feedback In Case of Malfunction

        GLP/SOP Should Describe Good Performance and Trouble
        Shooting/Corrective Actions

QUALITY CIRCLES

What They Are

    Small Groups with Kindred Interests/Responsibilities
    Meet at Regular Intervals on Company Time
    Meet With or Without Discussion Leaders

What They Do

    Think Quality
    Think About Sources of Error
    Identify Problems/Potential or Real
    Solve or Seek Help to Solve Problems
    Suggest Remedial Actions
    Preventive Maintenance

Benefits

    Grass Roots Input
    Increase Morale
    Promote Teamwork
    Create Problem-Prevention Attitudes
    Solutions Gain Acceptance Resulting from Concensus Approach
    Boost Quality/Productivity

QUALITY CIRCLE DISCUSSION TOPICS

Program/Project Management
- Problem Identification
- Defining Project Objectives
- Experimental Design
- Simple Models
- Factorial Design
- Project Organization

Quality Assurance Program Development
- General Aspects
- Chain of Custody
- OA Responsibilities
- Whats Right/Wrong with our QA Program

Quality Control Techniques
- Good Laboratory Measurement Practices
- Methods of Sample Preparation
- Contamination Control
- Analytical Factors Influencing Data
- Control Charts

Quality Assessment Techniques
- Quality Assessment Samples
- SRM's
- Collaborative Tests
- Plotting Data
- Inspection for Quality
- Performance Audits
- System Audits

Statistical Techniques
- Precision/Accuracy Concepts
- Regression Techniques
- Fitting Equations to Data
- Statistical Tests
- Analysis of Variance
- Statistical Reporting of Data

General
- Nomenclature
- Definitions
- Critical Consideration of Specific Techniques
- Sampling
- Maintenance of Equipment
- Good Housekeeping
- Book Reviews

Safety
- Basic Laboratory Precautions
- Chemical Hazards
- Physical Hazards
- Waste Disposal
- Storage of Chemicals
- Space Considerations

NOTE: The above topics can also provide the basis for a short course to indoctrinate new employees.

MANAGEMENT RELATED PROBLEMS

o   Inadequate Inspection/Evaluation

   Measurer
   Supervisor

o   Undefined/Illdefined Limits
o   Corrective Actions Illdefined or not Followed
o   Multiple/Simultaneous System Changes
o   Inadequate Training or Trainee Evaluation
o   Failure to Follow Instructions or Initiation of Unauthorized Procedural Changes

o   Workload Pressures
o   Quality Assurance Treated as a Step-Child

Adapted, in part, from A. Hainline, Jr., <u>Laboratory Management</u>, pp. 27-9 (October 1974).

QUALITY CIRCLES
John K. Taylor

ORGANIZATION

A quality circle consists of a small group (ten to twelve maximum) of employees with similar interests/involvements in an activity/process which may benefit from quality improvements. Since almost any activity/process should be concerned with the quality of its outputs, there is virtually no situation for which a Circle is not applicable. The group should be reasonably homogenous but not identical in its interest/involvement to provide a sound basis for attack of problems.

An organization planning for several quality circles usually will appoint a facilitator who will coordinate and direct circle activities.

Each circle will have a leader, usually appointed, but possibly elected by the group. The leader is responsible for the smooth operation of the Circle and must involve the participation of all members. Participation of the quiet members is encouraged by asking questions, seeking opinions, etc. Over-participation of the exuberant is discouraged by the idea-writing approach which will be discussed later.

OBJECTIVES

The objective of a Circle is both to prevent and solve problems related to quality of output. While their own outputs are of major concern, outputs of others related to theirs can also be considered. There is also the possibility of colaboration with others on such problems. The idea is to reduce errors and to enhance quality. Cooperation is also encouraged along with participation and motivation. Encouragement of ownership of change and grass roots inputs are additional objectives. Circles can become an organizational resource, consisting of teams experienced in trouble-shooting. But more than this, they are an excellent way to train. Circles, by virtue of an intensive look at a measurement process can provide a mechanism for education on that process (and even research in some instances). Thus a serendipic benefit is improved understanding of the process.

OPERATION

The recommended mode of operation for a Quality Circle is shown diagramatically in the figure. The mode is the same whether the objective is to identify problems to solve or to solve problems already identified.

In the case of both, an _activity_ or a _process_ is studied to decide the most beneficial course of action. This may require _study_ and often will require critical consideration to _define_ the problem area more clearly.

Ordinarily several aspects of the problem will be identified, in which case the <u>options</u> need to be generated. From the thinking of the Circle, a number of ideas will surface and <u>idea generation</u> should be encouraged. Even trivial and loosely related ideas should be accepted since these, if not directly useful, may stimulate other ideas, thus broadening the basis for selection and action. The recommended procedure is via <u>idea writing</u> which minimizes the direction of thinking by dominant individuals.

The <u>written ideas</u> are <u>collected</u> in an idea box, for example (writer's name not included). After reading by each member, the writing process may be continued for additional cycles to the point of diminishing returns.

The <u>selection</u> process consists of two activities: <u>categorization</u>; and <u>prioritization</u>. Categorization, or clustering consists in grouping by similarities i.e. to go from "local" to "global". This is by way of discussion, in which recorded ideas are classified by group concensus (but not rejected). During the discussion, the ideas may be clarified and edited, as needed. The process can be carried out by <u>intercomparison</u>, i.e. to distinguish between similarities and differences, on a one-to-one basis. This process will result in several <u>groups (clusters)</u> with similar characteristics, the general characteristic of which can be identified. While all clusters will have merit, some will be perceived to be more urgent (or important) than others. <u>Prioritization</u> can then be established by group preferential rating.

The actions that should be taken will fall into three classes:

> Iteration - further study may be required to define a problem more clearly for specific action; the problem may need collaborative effort with other circles/groups.

> Implementation - the circle may have the authority to act on some of its own recommendations with little or no approval required by others.

> Recommendation - most problem solving will fall in this class in that the solution may need approval by management or interaction with others which may require external approval. In such cases the circle will decide on the best course of action to follow in presenting its recommendation to management.

PRESENTATION

Recommendations may go to management by two routes. <u>Written reports</u> provide a record of a circle's outputs and are always needed. In addition, <u>oral presentations</u> may be made, as well. The circle should be well-prepared in such a situation, calling for dry runs. If a dry-run does not appear to go well, it may mean that something is lacking in the subject matter. Hence, oral presentations, whether or not used as a final means of communication, are useful for deciding the merits of any recommendations made by a circle.

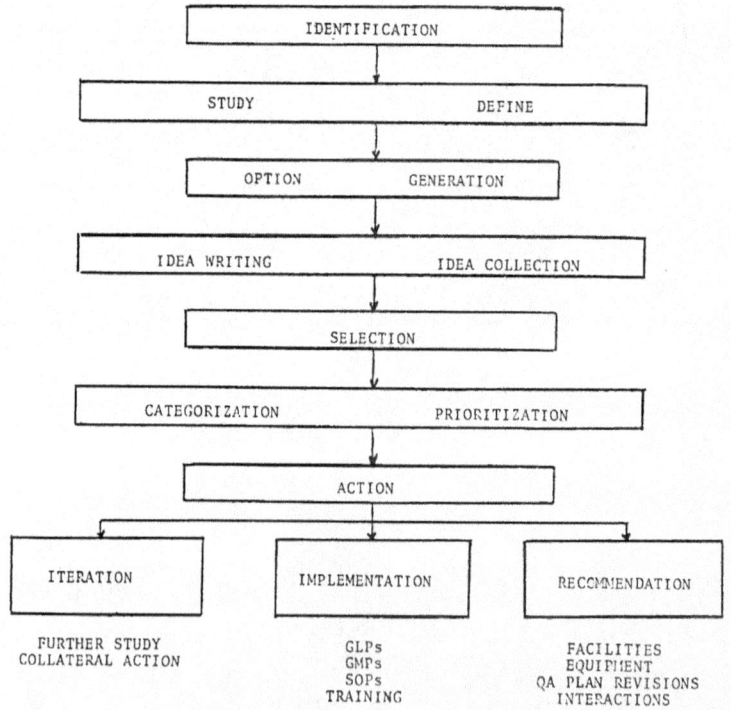

XIII

MEASUREMENT COMPATIBILITY

THE NATIONAL MEASUREMENT SYSTEM

TRACEABILITY

THE TROUBLE WITH THE IDEA OF MEASUREMENT IS ITS SEEMING SIMPLICITY

Problem
                           o o o o o
    How many dots?
    What is the distance between them?

        Needed
            Unit of length
            What to measure
            Where to measure
            Measurement device
            What affects measurement?

        Conclusions
            Counting is exact
            Calculation can be as exact as one chooses
            Measurement is inexact

MEASUREMENT PROCESS

Conceptual Foundation
- Phenomenon
- Definitions
- Concepts
- Quantities
- Units

Basic Technical Infrastructure
- Knowledge
- Documentation
- Specifications
- Reference Data
- Reference Materials

Realized Measurement Capabilities
- Instruments
- Skill
- Ranges
- Precision
- Accuracy

Dissemination/Enforcement Network
- Education/Professional Societies
- Standards/Testing Laboratories
- Regulatory Agencies
- States Weights & Measures Labs
- NBS

End-Use Measurements
- Process Control
- Commerce
- Evaluate Materials
- Develop New Technology
- Accumulate Knowledge

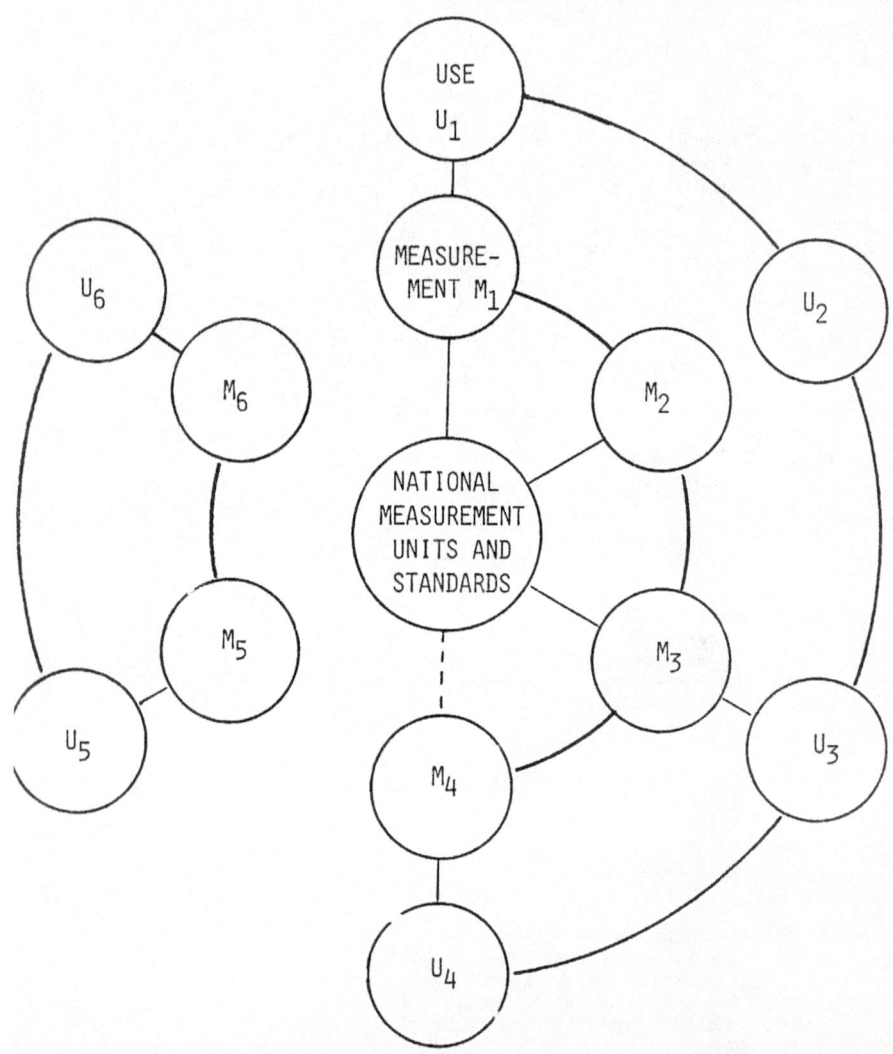

TYPICAL MEASUREMENT SITUATIONS

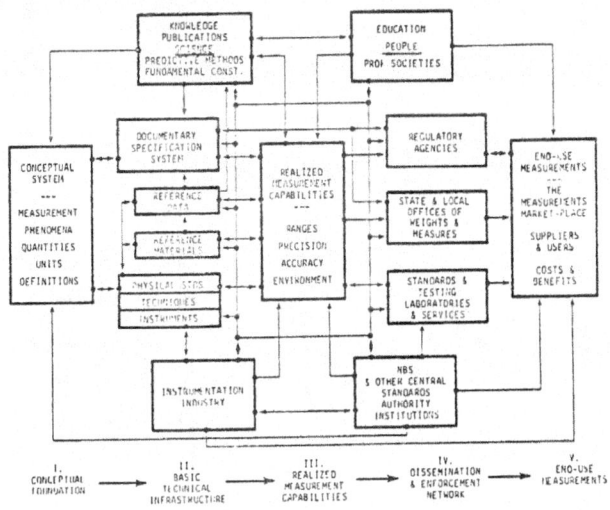

FLOW DIAGRAM OF THE NATIONAL MEASUREMENT SYSTEM

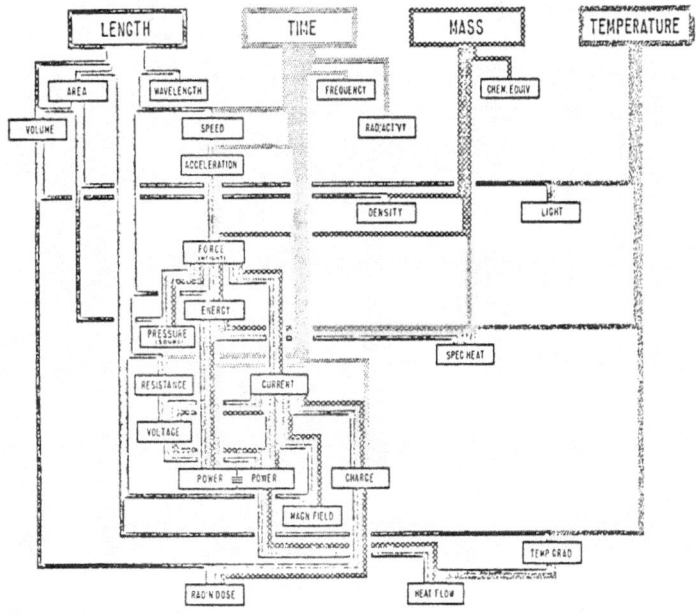

GENEALOGY OF A MEASURING SYSTEM
(RELATIONSHIPS SHOWN ARE TYPICAL. MANY IMPORTANT QUANTITIES SUCH AS ANGLE, INDUCTANCE, CAPACITY, VISCOSITY, ETC., ARE OMITTED TO KEEP CHART SIMPLE.)

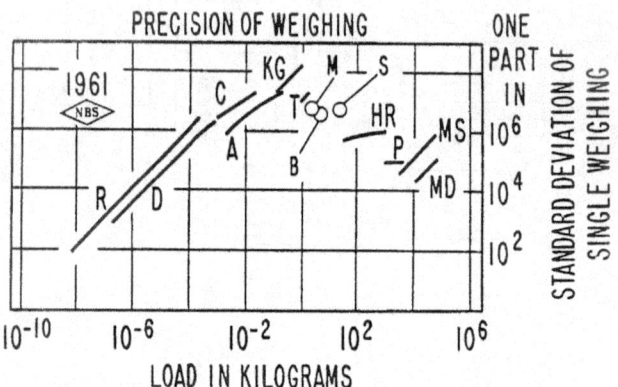

T 2 kg BALANCE (NBS T-1)
M 6-lb BALANCE
B 10 kg BALANCE (NBS B-1)
S {25 kg BALANCE (NBS S-1)
    {50-lb BALANCE
R QTZ-FIB ULTRA-MICROBALANCE
D ASSAY BALANCE
C CORWIN BALANCE

A 200 g BALANCE (NBS A-1)
KG RUEPRECHT 1 kg BALANCE
HR RUSSELL BALANCE, 2.5 k-lb
P PLATFORM SCALE, 10 k-lb
MS MASTER SCALE, 150 k-lb, SUB-
    STITUTION WEIGHING
MD MASTER SCALE, DIRECT READING

PHYSICAL MEASUREMENTS

Characteristics

- o Ordinarily made on well characterized defined systems
- o Made by relatively simple measuring systems with limited number of sources of variance
- o Sources of variance readily identifiable
- o Systems stable or vary in a predictable manner
- o Limited problem of interferences
- o Quality assurance need recognized
- o Measurement assurance programs established
- o Calibration networks in place

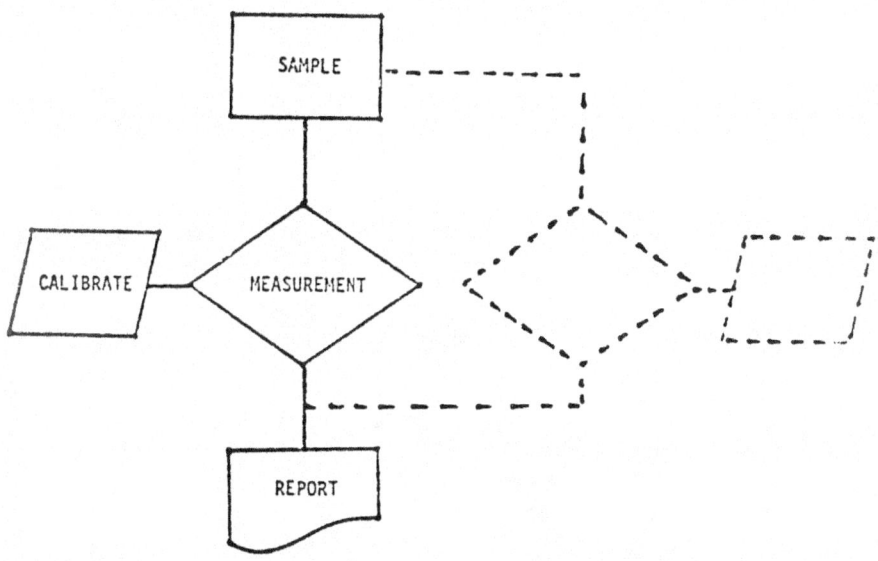

TYPICAL PHYSICAL MEASUREMENT

XIII-5

CHEMICAL MEASUREMENTS

Characteristics

- o Often made on ill-defined materials
- o Measurement errors may approach or exceed "product" tolerances
- o Made by complex, multicomponent measurement systems with multiple sources of variance
- o Difficult to identify and/or isolate sources of variance
- o Measurements often made on fugitive or rapidly-varying populations with little or no opportunity to check results
- o Users (and often practioners) fail to recognize complexities of problems
- o Measurements may be made with insufficient regard for quality assurance procedures

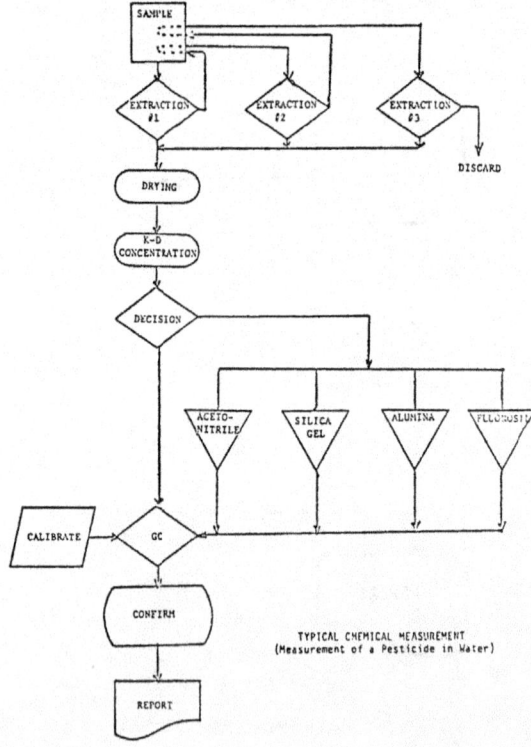

TYPICAL CHEMICAL MEASUREMENT
(Measurement of a Pesticide in Water)

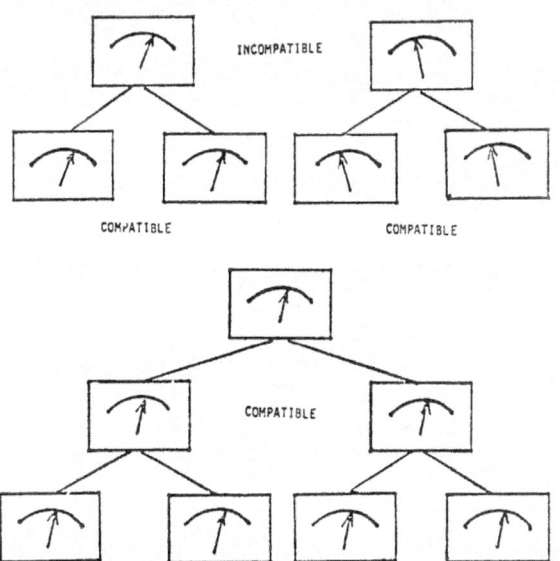

COMPATIBILITY IS WITH THE STATION THAT MEASURED THE MEASURAND

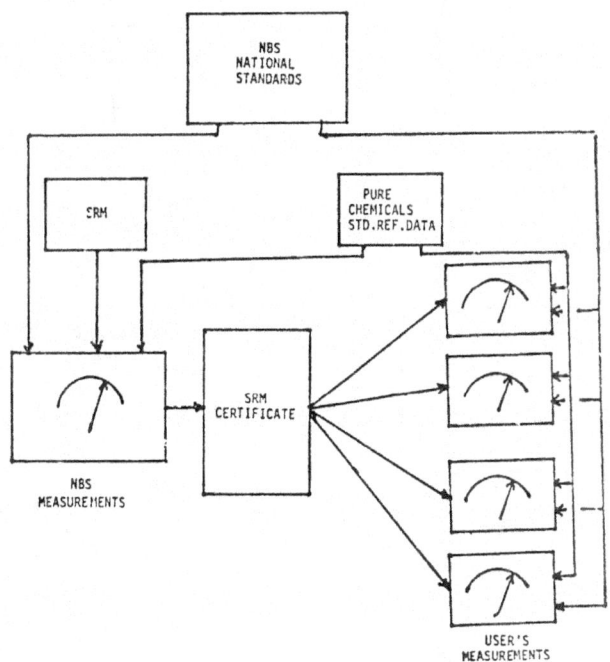

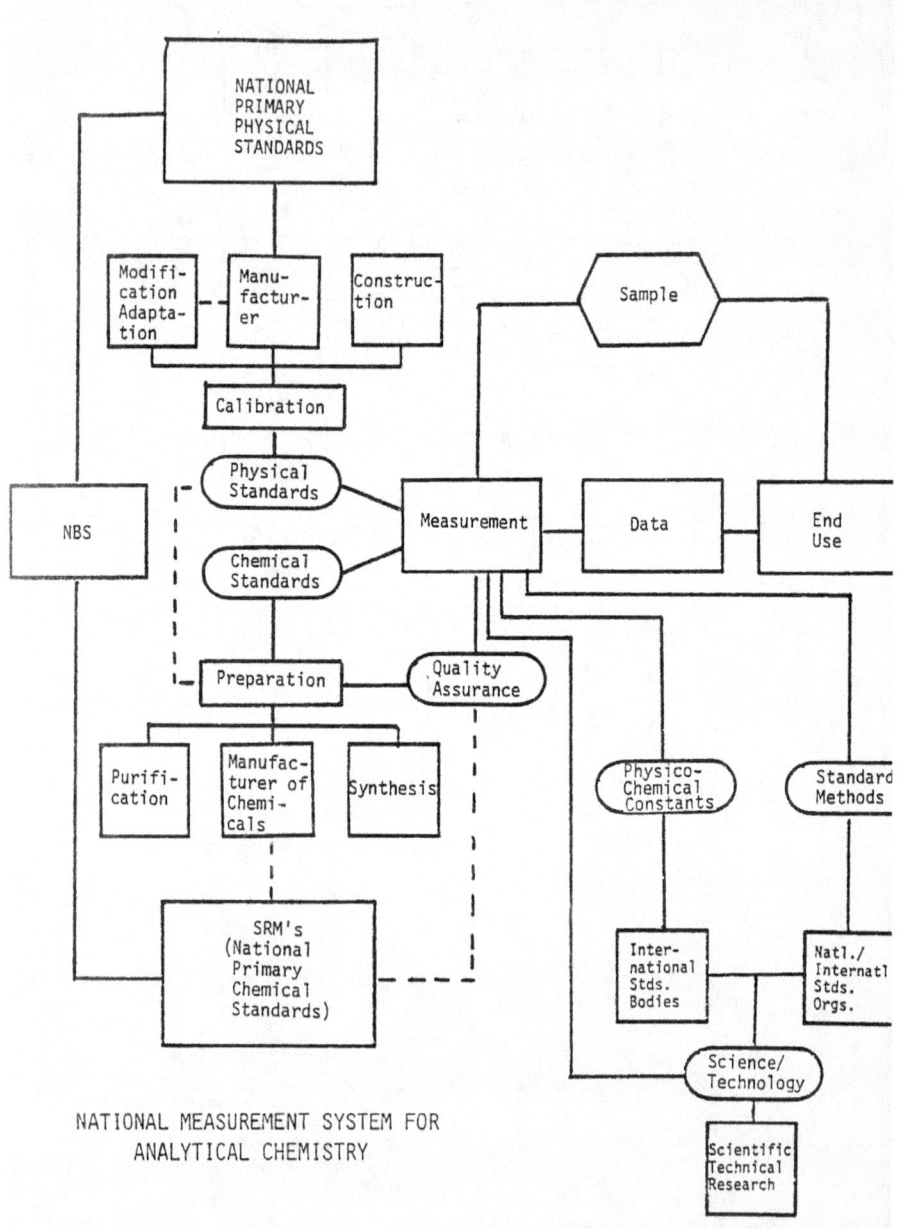

NATIONAL MEASUREMENT SYSTEM FOR
ANALYTICAL CHEMISTRY

TRACEABILITY

DEFINITIONS

Trace - To discover or uncover by going backwards over the evidence, step by step; to ascertain, establish, attribute as a result of such retracing or reviewing, as to trace the cause of an epidemic, one's descent from the pilgrims, etc.

Traceability n - Able to be traced.

Webster's New International
Dictionary, Second Edition

TRACEABILITY is GENEALOGY

Pathways must be defined
Uncertainties must be defined
Credibility depends on

    Validity of Pathway
    Reliability of Measurements

        Quality Assurance

    Ancillary Phenomena

        Stability
        Chain of Custody

TEN CARDINAL GUIDELINES FOR TRACEABILITY

1. A measurement system must be known to be in control to insure confidence in any measurement. The objective is to maximize confidence and minimize number of measurements.

2. Measurements essentially involve intercomparison of an unknown with a standard and control can be demonstrated by replicates on the standard, the unknown, or a combination of such measurements.

3. Quality control procedures may pool measurements to interrelate them on a time-sequence so that any particular measurement is supported by others made at the same or different times.

4. When a measurement standard has been certified by NBS, intercomparisons with that standard constitutes direct traceability, with the uncertainty of the measurement process.

5. When a measurement is made with respect to a secondary standard certified with respect to an NBS standard, indirect traceability may exist within uncertainty of the various measurements.

6. Comparison of a measured property with that of a certified material may provide traceability by inference, depending on the validity of the inference.

7. Only the property measured is traceable. Any property inferred from such a measurement must be supported by evidence, which may not be traceable.

8. In any traceability situation, NBS responsibility is only for the standards it certifies. The Measurement Laboratory has the responsibility for its own measurements and all claims related to them.

9. Any measurement is valid only at the time of measurement. Any extension of the measured value on a time basis must be supported by other evidence.

10. Any measurement is only as reliable as the measurer.

XIV

REFERENCE MATERIALS

NOMENCLATURE

Reference Material (RM) - Substance of which one or more properties are established for use to calibrate or verify a measurement.

Internal Reference Material (IRM) - A reference material developed by a laboratory for its own internal use.

External Reference Material (ERM) - A reference material provided by someone other than the end-user laboratory.

Certified Reference Material (CRM) - A reference material accompanied by a certificate issued by an organization generally accepted as technically capable to do so.

Standard Reference Material (SRM) - National Bureau of Standards certified reference material.

CLASSES OF REFERENCE MATERIALS

Grade A - Atomic weight standard

Grade B - Ultimate standard - a substance which can be purified to virtually Grade A.

Grade C - Primary standard - commercially purified to a purity of 100 ± 0.02%.

Grade D - Working standard - commercially available, purity of 100 ± 0.05%.

Grade E - Secondary standard - of lower purity, standardized against Grade C material.

Reagent Water - Not defined as above. ACS and ASTM have specifications.

ASTM D1193 - Reagent Water

ASTM E200 - Standard Solutions, Preparation Standardization, Storage

Reagent Chemicals - Sixth Edition, (1981) American Chemical Society, 1155 16th St. N.W., Washington, DC 20016

USE OF REFERENCE MATERIALS

Used to Assess Accuracy of

Systems in Statistical Control

Reference Material must be

Identical to or

Simulate Test Material

Interpretation of Reference

Material Data is by Inference

ADDITIONAL USES OF RM'S

To Validate Test Methods for a Specific Use

To Test Proficiency of Analyst

To Measure Performance of Methods Under Development

    IRM's - Monitor <u>Precision</u>; <u>Accuracy</u> in Special Cases

    SRM's - Monitor <u>Accuracy</u>; <u>Precision</u> in Special Cases

Direct Traceability

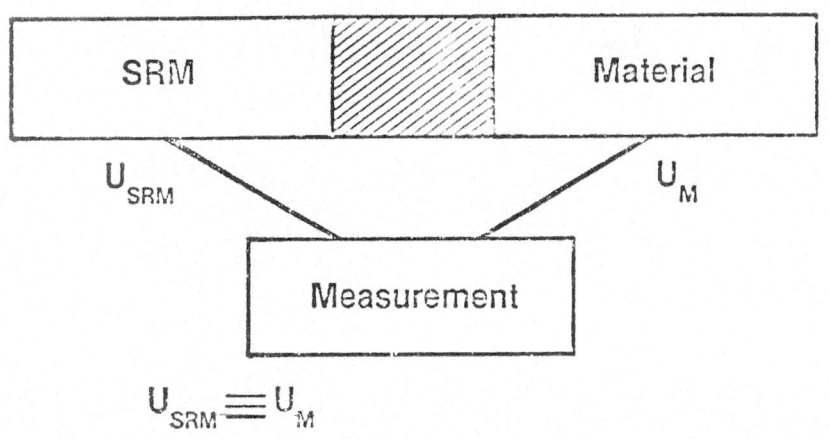

Indirect Traceability
Case 1

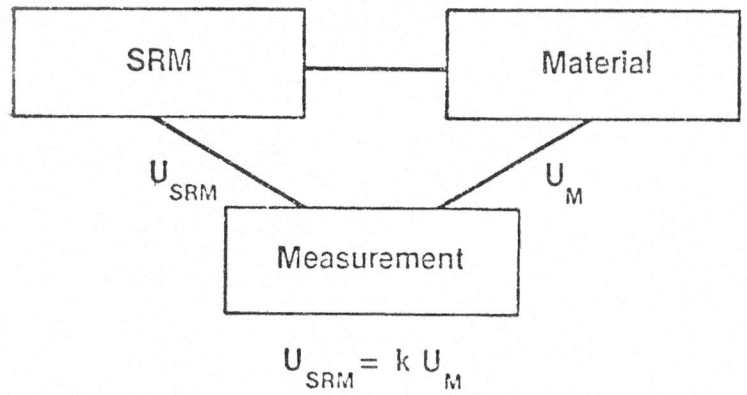

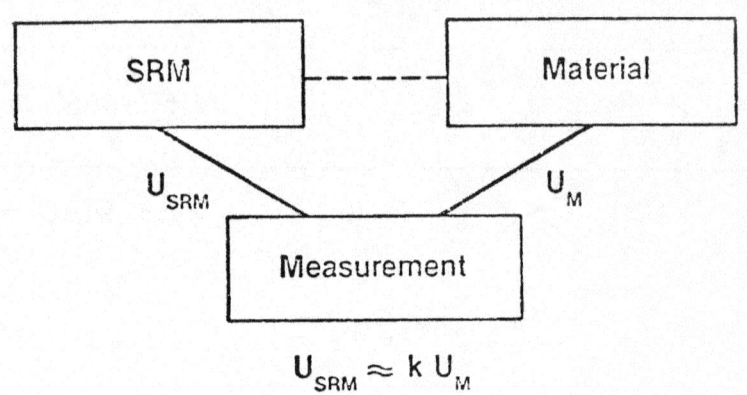

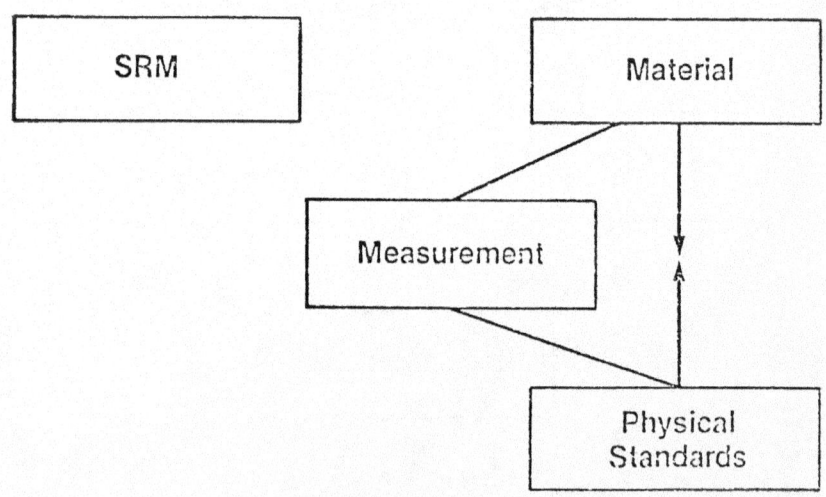

SRM

Well-Characterized Material For:
    Calibration of Measurement System
    Production of Scientific Data
    Basis of Measurement Comparison
    Control of Production Process

Material Must Have:
    Homogeneity
    Stability
    Availability

SRM

    Actual
    Synthetic
    Simulated
    Foreign

STEPS TO DEVLEOP AN SRM

Establish Need
Develop Material
Develop Measurement Method
Study Stability of Material
Obtain SRM Material
Process SRM Material if Necessary
Measure Homogeneity
Measure for Certification
Package Material
Prepare Certificate
Announce Its Availability
Distribute Material
Do Follow-up Studies to Insure Its Reliability

CERTIFICATION

Measure Homogeneity by at Least One Method
Measure Property by Either
    o   Well-Established Reference Method
          or
    o   At Least Two Independent Techniques
Both made by Pre-Established Measurement Plan
Resolve all Problems
Statistically Analyze Data
Establish Confidence Limits
Review all Data before Release
Retain Records for Future Use

PROBABLY MORE NONSENSE IS TALKED ABOUT MEASUREMENT THAN ABOUT ANY OTHER PART OF PHYSICS (OR CHEMISTRY).

MEASUREMENT DOES NOT PRODUCE A VALUE, BUT A RANGE OF VALUES

THE REAL UNCERTAINTY OF ANY MEASUREMENT IS NOT WHOLLY THE ERROR COMMITTED IN MAKING THE FINAL MEASUREMENT, THAT IS TO SAY THE FINAL SCALE READING.

TWO ATTEMPTS AT MEASURING THE SAME THING WILL PRODUCE TWO RANGES THAT MAY NOT OVERLAP AND WE CANNOT BE SURE THAT THE TRUE VALUE LIES WITHIN EITHER OR BOTH OF THEM.

OFTEN THE TRUE ERROR IS SO LARGE THAT THE ERROR OF THE FINAL READING MAY BE IGNORED.

LET US KEEP SIGNIFICANT FIGURES WHERE THEY BE LONG, AS A CONVENIENT CONVENTION FOR WRITING ANSWERS IN PURE ARITHMETIC. THEY HAVE NO OTHER USE.

D. P. DELURY - "COMPUTATION WITH APPROXIMATE NUMBERS", NBS SP 300, p. 392 (1969).

XV

REPORTING DATA

NO MEASUREMENT IS SCIENTIFIC
UNLESS IT'S UNCERTAINTY IS KNOWN

NO MEASURED VALUE IS SCIENTIFICALLY
STATED UNLESS IT'S UNCERTAINTY IS
EXPLICITLY STATED

Error bars must be associated with every data point so that the strengths and weaknesses of every decision based thereon may be evident. This will support valid interpretation and prevent overinterpretation of data.

MINIMUM REQUIREMENTS FOR REPORTING DATA

Sample Documentation
- Methodology
- Number
- Location - Time - Space

Measurement Documentation
- Methodology
- Calibration
- Quality Control/Quality Assessment
- Intercalibration
- Precision
- Accuracy

Data Documentation
- Confidence Limits
- Measurement
- Sample

DATA LIMITATIONS

Precision and Accuracy
Confirmation

    Independent Method
    Independent Conditions

Recovery

    Natural vs. Spiked Samples
    Interpretation

Limit of Detection
Limit of Quantitation

UNCERTAINTY LIMITS

One Approach

    Use of Significant Figures

        Round off so that only last figure is uncertain, i.e., one-half unit in last figure reported or ± 5 units in next unreported place.

        Problem:

            What is Uncertain?
            Need Rules to Decide.

Better Approach

    Use of Statistical Limits

        Measurement Precision

        Bounds to Systematic Error

            State in Sentence Format
            State Separately

        Report Above Limits to Two Figures
        Round Off Data to be Consistent with the Limits

STATISTICAL CONFIDENCE LIMITS

Width of Confidence Interval for the Mean

$$\bar{x} \pm \frac{z\sigma}{\sqrt{n}} \qquad \bar{x} \pm \frac{ts}{\sqrt{n}}$$

Statistical Tolerance Limits

For Fixed Percentage of Population

$$\bar{x} - Ks \text{ to } \bar{x} + Ks$$

Bounds to Systematic Error

Determination of Reasonable Limits
Involves an Element of Judgment
Limits Cannot be Set in Exactitude

BOUNDS FOR BIAS

Physical
    Factors that Affect Measurement
    Factors that Affect Samples

Chemical
    Inteferences
    Blanks
    Calibration

Correction for Known Sources of Error
    Error analysis made; corrections evaluated and applied
    All corrections adequately described

Documentation
    All of above documented by:
        Experimental Data
        Literature References
        Detailed Explanations

ASSIGNED LIMITS OF UNCERTAINTY

Based upon standard deviation or its estimate, and number of measurements of analate of concern

Must distinguish between measurement and sample variance in most cases

Unacceptable limits may be reduced by:

    Improving measurement precision
    Improving sample homogeneity (also compositing, etc.)
    Increasing number of measurements
    Increasing number of samples

Cost/Benefit Considerations Involved in Decisions of Acceptability

CONFIDENCE/TOLERANCE INTERVALS

df/n	$t_{.025}$	$\dfrac{t}{\sqrt{n}}$	$K_{95/95}$
1	12.706	8.954	---
2	4.303	2.484	37.674
3	3.182	1.591	9.916
4	2.776	1.241	6.370
5	2.571	1.050	5.079
6	2.447	0.925	4.414
7	2.365	0.893	4.007
8	2.306	0.769	3.732
9	2.262	0.715	3.532
10	2.228	0.672	3.379
15	2.135	0.533	2.954
20	2.086	0.455	2.752
25	2.060	0.403	2.631
30	2.042	0.367	2.549
100	1.99	0.199	2.233
∞	1.960	0	1.960

95% Confidence Interval of Mean

$\dfrac{t}{\sqrt{n}} \cdot s$ or 1.96σ

Statistical Tolerance Interval Ks

CONFIDENCE/TOLERANCE

Intervals of a Mean
(Unit Standard Deviation)

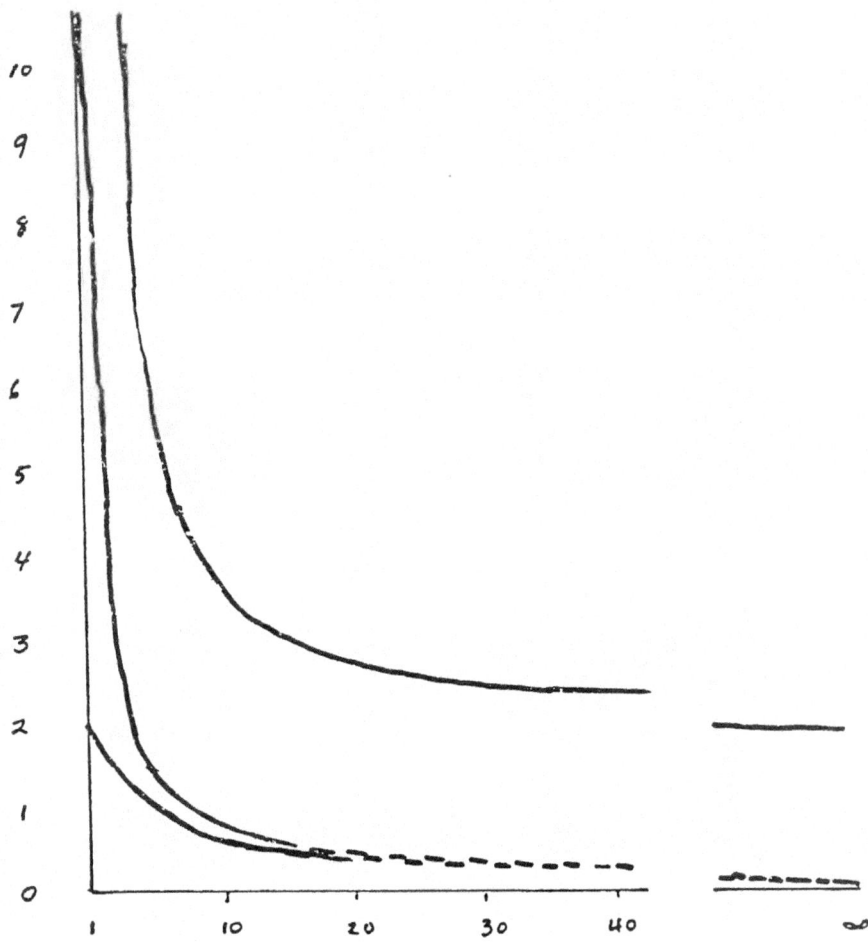

1 - Tolerance Limits

2 - Confidence Limits of Mean, Based on s

3 - Confidence Limits of Mean, Based on σ

REPORTING NUMERICAL RESULTS

Calculate All Means To At Least One More Significant Figure Than Is In Data

Calculate Standard Deviation Estimates To At Least Two Significant Figures

Calculate Confidence Interval and then Round Off to Two Significant Figures

Round Off Mean Consistent with Confidence Interval

Report Mean and Its Confidence Interval, Stating What It Is

<u>Don't Round-Off Too Early</u>. Keep As Many Figures As Possible When Recording Data and In Calculations, Let the Data Decide Its Significance

ANALYTICAL REPORT

Title
Client
Problem

    Objectives
    What, Why Done

Content

    Sample(s)
        Identification
        History
        Serial Numbers
            Client
            Laboratory

    Description of What Was Done
        Procedure
        Methodology
            Reference or Description
        Data
            Summary, Uncertainty
            Reference to Laboratory Records (May be Blind)
        Interpretation(s)
            With Respect to Problem
        Recommendation(s)

Attestation

    Analyst; Supervisor; Management

---

* (Some of this can be the reference.)

ACS GUIDELINES FOR DATA ACQUISITION AND DATA QUALITY
EVALUATION IN ENVIRONMENTAL CHEMISTRY

Anal. Chem. $\underline{52}$; 2242-49 (1980)

PRINCIPLES OF ENVIRONMENTAL ANALYSIS,

Anal. Chem., $\underline{55}$, 2210 (1983)

SUMMARY

o Well-designed, Carefully Executed Measurement Process
o Use of
    Sensitive, specific, validated measurement methods
o Use of Reliable Protocols for
    Sampling
    Measurement
o Use of Quality Assurance Procedures
    Demonstration of Statistical Control via. Control Charts
o Validation of
    Samples
    Measurements/Data
o Assignment of Uncertainties to Data
    Measurement/Sample
o Specifics:
    LOD, LOQ, Qualitative Confirmation
    Recovery Verification
    Use of Reference Materials

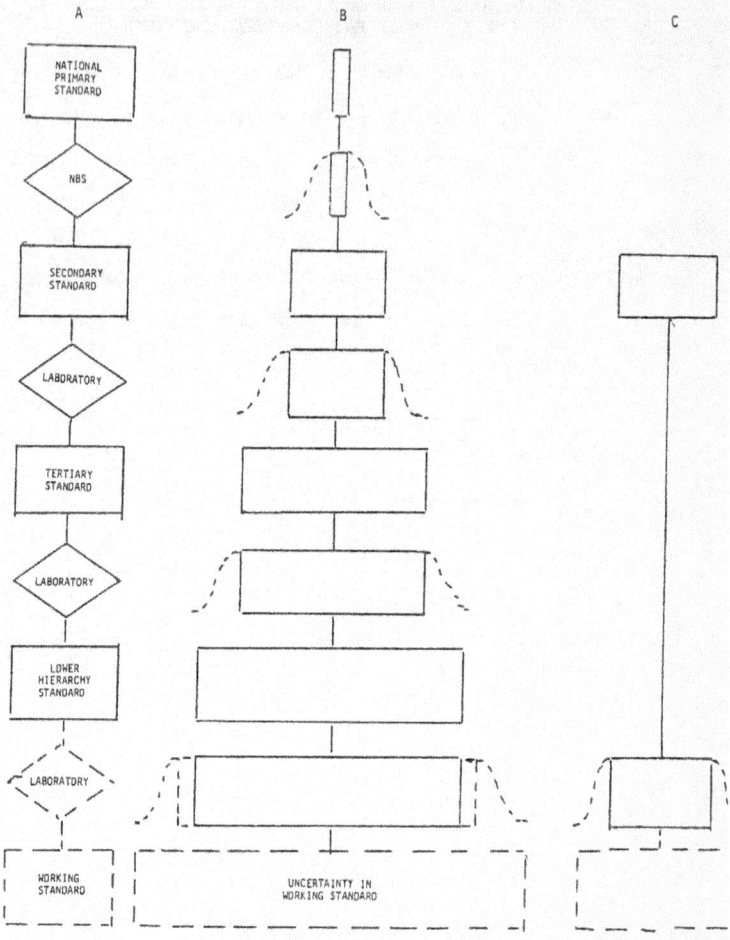

A. CALIBRATION CHAIN

B. PROPAGATION OF UNCERTAINTY

C. ALTERNATE CALIBRATION OF CHAIN UNCERTAINTY

FIGURE   PROPAGATION OF CALIBRATION UNCERTAINTY

XVI

VALIDATION

VALIDATION

Valid (definition)

> Founded on Truth or Fact
> Capable of Being Justified, Supported, or Defended
> Not Weak or Defective
> Well Grounded
> Sound
> Accomplishing What is Claimed or Intended
> Supported by Truth
> Having Legal Efficacy or Force

Validation (definition)

> The Act or Process of Validating
> (See Appendix D)

SAMPLE VALIDATION

Purpose

> To Accept Individual as a Member of Population Under Study
> To Admit Sample for Analytical Measurement
> To Minimize Later Questions on Sample Authenticity
> To Provide Opportunity for Resampling When Needed

Criteria for Acceptance

> Positive Identification
> Meets Physical/Chemical Specifications
> Valid Chain of Custody

Criteria for Rejection

> Sampling System Not in Control
> Erroneous/Conflicting Data on Identity/Character
> Questions on Sample that Cannot be Removed or Clarified
> Sample Cannot be Unequivocally Considered Member of Population

A VALID MEASUREMENT PROCESS

Should Produce:

> A Useful Measured Value
> An Estimate of the Uncertainty of that Value

A Reported Value

Should Include:

> Best Estimate of the Value, and
> Its Estimated Uncertainty

WHAT IS A VALID METHOD?

Performance Parameters Acceptable
    Definition of Acceptable Limits must Precede Selection
Required Performance Parameters
    Detectability
    Sensitivity
    Selectivity
Accuracy Adequate
    Precision
    Bias
Experimental Demonstration of Above
    Using Evaluation Samples Equivalent to Test Samples

OTHER CONSIDERATIONS

Interefences
Cost
Availability of
    Equipment
    Experience
Sample Load
Calibration
Skill Requirements
Portability
Time Requirements
Ruggedness
Down Time
Other Options

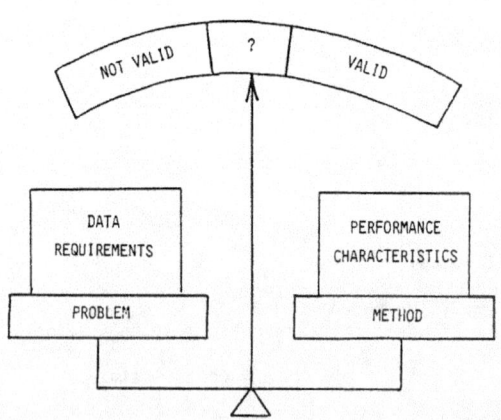

VALIDATION PROCESS

General Validation

    Technique - Research of Scientific Community
    Method - Research of Individual Scientists; Applied Research
    Procedure - Users; Standardization Organizations

    Protocol - Fiat

        Based on Decision Process

Result

    Demonstration of General Validity
    Definition of Areas of Applicability
    Delineation of Some Specific Uses

        METHODOLOGY MAY BE DEVELOPED:

For General Purposes
For a Class of Uses
For a Highly Specific Use

If carefully and properly done, a method which is valid for some purpose should result.

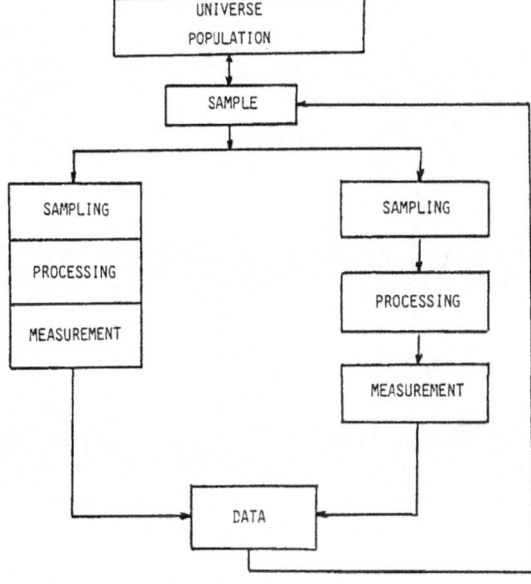

END-USE VALIDATION

User Must Demonstrate/Confirm Validity for a Specific Use by:

- Use of "Identical" Reference Samples
- Use of "Analogous" Reference Samples
- Comparison with a Known Trusted Method
- Independent Method Confirmation
- Spikes/Surrogates

Demonstration of Attainment of Statistical Control Mandatory in Each of Above Cases.

DATA VALIDATION

(The Bottom Line)

Valid Method Necessary But Not Sufficient

Other Requirements

Demonstration of Competence of Analyst
Demonstration of Valid and Operational Quality Assurance Plan
Attainment of State of Statistical Control

Together With

Validity of Model
Validity of Sample

DATA VALIDATION

The process whereby data are filtered and accepted or rejected, based on a set of criteria.

A systematic procedure of reviewing a body of data against a set of criteria to provide assurance of its validity prior to its intended use.

Data Validation is

After the Fact
Applied to Body of Data
Systematically and Uniformly Applied

Data Validation Must be

Close to its Origin
Independent
Objective

Criteria

Checks for Internal Consistency
Checks for Temporal and Spatial Consistency
Checks for Proper Identification
Checks for Transmittal Errors
Checks for Blunders

Techniques

Intercomparisons
Reasonableness

    vs. A Priori Limits
    vs. A Posteriori Limits

Data Plots
Regression Analysis
Test for Outliers

    Data Flagged or Rejected

VALID DATA

Raw Data

    Produced By a Valid Process

        Sample
        Method
        Calibration
        Quality Assurance

Finished Data

    Screened for Consistency
    Elimination of Outliers
    (to the Extent Possible)

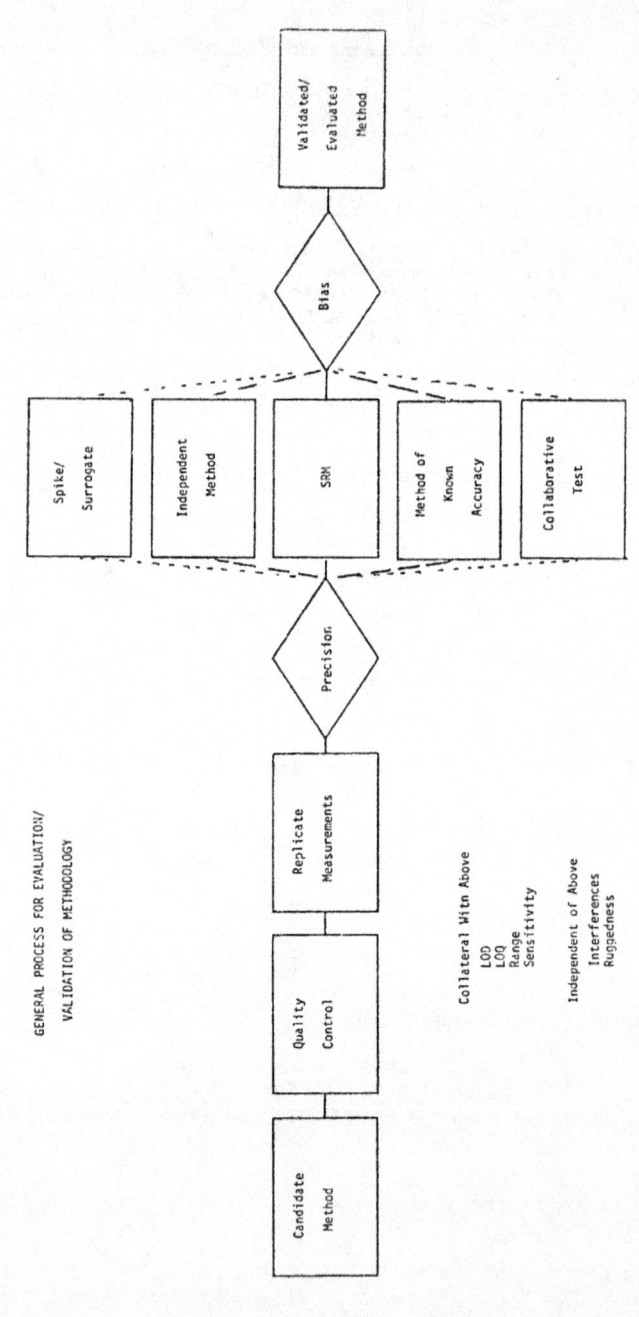

GENERAL PROCESS FOR EVALUATION/
VALIDATION OF METHODOLOGY

XVI-6

XVII

LABORATORY

CERTIFICATION/EVALUATION

CRITERIA FOR EVALUATING LABORATORIES

ASTM E-548   Generic Criteria for Use in the Evaluation of Testing and Inspection Agencies (6)

ASTM D-3856   Evaluating Laboratories Engaged in Sampling and Analysis of Water and Waste Water (3)

ASTM D-3614   Evaluating Laboratories Engaged in Sampling and Analysis of Atmospheres and Emissions (2)

ISO Guide 25   Guidelines for Assessing the Technical Competence of Testing Laboratories (23)

ASTM E-743   Spectrochemical Laboratory Quality Assurance (7)

ACIL   Quality Control Systems Requirements (38)

LABORATORY ACCREDITATION

Evaluation of the Capability of a Testing/Inspection Laboratory in Specific
    Fields of Activity

        U.S. Accreditation Systems

            44 Private
            26 Governmental

            Some Voluntary
            Some Mandatory

    Benefits

        Increased Confidence/Acceptance
        Identification of Competent Labs
        Focus on Good Practices
        Provide Goals and Criteria

    Disadvantages

        Retards/Freezes Technology
        Discourages Innovation/Initiative
        Petty Annoyances

    Costs

        Fees
        Administration
        Questionnaires
        On-Site Inspections
        Proficiency Testing

BASIC SYSTEMS

Product Focus

    Ability to Test Products (Specified) Using Specified Technology (e.g., NVLAP)

Discipline Focus

    Ability to Use Test Technology in Specified Test Areas (e.g., AALA, SCC NATP)

Similarities

    Both Set Criteria
    Both Evaluate
    Both Accredit

Differences

    P-F Highly Specific
    D-F Considerable Generality
    P-F Requires Multiple Accreditation
    D-F Requires Only Multi-Area (Discipline) Accreditation

    AALA - American Association for Laboratory Accreditation
    NATA - National Association of Testing Authorities (Australia/New Zealand)
    NVLAP - National Voluntary Laboratory Accreditation Program
    SCC - Standards Council of Canada

PRODUCT FOCUS

    Identifies Needs
    Identifies Product
    Identifies Test Method(s)
    Identifies Criteria

    Test Specific

        Human Resources
        Space/Equipment
        Quality Assurance

    On-Site Inspection
    Proficiency Testing

        Major Factor
        Continuing Activity

DISCIPLINE FOCUS

Acts on Individual Requests
Identifies Disciplines
Identifies Criteria
- Discipline Oriented
  - Organization
  - Human Resources
  - Space/Equipment
  - Method Manuals (Emphasis on use of Standard Methods where possible)
  - Sample Control
  - Data Control
  - Traceability
  - Quality Assurance
  - Documentation
  - Report Requirements

On-Site Inspections
- Comprehensive - General

Proficiency Testing
- Minor

PRINCIPLES IN COMMON

Organization
- Well Organized
- Duties/Responsibilities Defined
- Supervision/Inspection/Audit/Self Appraisal

Staff
- Technical Competence
- Qualifications Documented
- Training/Maintenance of Competence
- Sufficient Supervision
- Adequate Support

Equipment
- Adequate in Kind/Quality
- Maintained

Calibration/Reference Standards
Test Methods/Procedures
Environment
- Space
- Physical/Chemical Control
- Housekeeping

Test Items
- Handling
- Storage
- Chain of Custody

Records
Test Reports
Quality Assurance Program

SELF APPRAISAL

Qualifications of Staff
    Education
    Experience - General
    Experience - Specific

Facilities and Equipment
Methodology
Performance
    A Laboratory Offering Services Should Have:

        Experience
        Methodology
        Facilities
        Equipment - Standards
        Knowledgable Personnel

        ALL SHOULD BE A MATTER OF RECORD AND DEMONSTRATED

RECOMMENDED MINIMUM REQUIREMENTS FOR PERSONNEL

General Management
    General Knowledge Consistent with Policy Decisions

Technical Director
    BS Degree
    License or Certificate as Required
    Five Years Experience in One or More Fields He Directs
    Affiliations with Technical/Professional Societies

Technical Supervisor
    BS Degree
    Five Years General, Two Years Special Experience
    Affiliations with Technical/Professional Societies Pertinent to
     Field
    Up-To-Date Knowledge of Field

Scientific Staff
    BS Degree
    On Job Training
    Demonstrated Competence

Technical Staff
    H.S./Some Technical Training
    On Job Training
    Familiarity with Test Methods

Support Staff
    General Competence

Consultant(s)

XVIII

PLANNING

QUALITY ASSURANCE

PROGRAMS

It is axiomatic that if you need an outsider to tell you that you have a quality problem, you have a problem.

QUALITY ASSURANCE
FACT OR FICTION
SENSE OR NONSENSE

Common Comments

"Its only common sense."
"If you don't care, nothing will happen."
"If you do care, you don't need it."
"It only works for routine situations"

"We always do special work."
"We do research."

"Unnecessary"

"Its only needed when you have lots of technicians."
"Just hire good people, don't bug them, and they'll turn out good work."
"We've been in this business for years without any quality problems."
"We subscribe to, and informally practice Q.A. There's no need to formalize."

PRODUCTION OF QUALITY DATA
is a COMMON GOAL

BUT

IT IS PURSUED DIFFERENTLY

- In various organizations
- Within laboratories of the same organization
- Within groups in the same laboratory
- From time-to-time by a given individual

PREMISE

A commonly accepted set of guidelines will promote greater uniformity

and

Harmonize on-going practices
Enhance compatibility of data

COST/BENEFITS OF QUALITY ASSURANCE

Costs

Direct
    Test Materials
    Standards
    Quality Assurance Equipment - Test Instruments
    Analysis of Quality Assurance Samples

        Time of Personnel
        Time of Supervision

    Quality Assurance Official
    Committee Work
    Round Robin Costs
    Travel/Attendance at Meetings

Indirect
    Training
    Extra Cost for Quality People
    Extra Quality Equipment
    Extra Quality Supplies
    Relaxed Work Schedules

Benefits

More Efficient Outputs
Less Replicates for Same Reliability
Less Do-Overs
Greater Confidence Of:
    Staff
    Laboratory
    Customers

INTANGIBLE

Benefits of a QA Program

- Promote External Image
- Improve Internal Image
- Promote Client Confidence
- Add Credence to Results
- Prevent Hasty Disclosures
- Minimize Indecision
- Eliminate Unnecessary Redundancies
- Promote Continuity of Effort
- Provide for Retention of Vital Records
- Set Forth Goals and Objectives
- Provide Guidance to Staff
- Provide Basis for Training

The nature of analytical work--production of numerical data--engenders expectation of objective evidence to demonstrate the degree of confidence in the results.

LEVELS OF QUALITY ASSURANCE

Individual
Laboratory
Consumer

> National
> International

QUALITY ASSURANCE

Responsiblities

Top Management

> Policy
> Climate
> Resources

Supervision

> Guidance
> Direction
> Surveilance

Individuals

> Quality of Outputs-Technical Competence

Quality Assurance Officer

> Consultation
> Advice
> Oversight

QUALITY ASSURANCE

Implementation

Policy

> Established by Top Management

Procedures

> Developed by Concensus Action of All Participants

Supervision

> Exercised by Management Chain

Oversight

> By Quality Assurance Officer, Reporting to Top Management, Independent of Normal Management Chain

QUALITY ASSURANCE HEIRARCHY

```
┌─────────────────┐
│                 │
│     PROGRAM     │
│                 │
└─────────────────┘

┌─────────────────┐
│     GLP'S       │
│     GMP'S       │
└─────────────────┘

┌─────────────────┐
│                 │
│     SOP'S       │
│                 │
└─────────────────┘

┌─────────────────┐
│                 │
│     PSP'S       │
│                 │
└─────────────────┘

┌─────────────────┐
│                 │
│ IMPLEMENTATION  │
│                 │
└─────────────────┘
```

WHO DOES WHAT

Management

    Decision to go QA
    Commits Resources
    Designates Leader
    Approves Appropriate Stages

Leadership

    Develops Plans
    Gets Cooperation/Involvement
    Gets Consensus Approval

Staff

    Provides Expertise
    Technical Advice/Guidance
    Writes or Reviews Appropriate Plans

QA PROGRAM/PLAN DEVELOPMENT

1. Identify Goals

    Motivation

        Internal
        External

2. Identify On-Going QA

    Describe in Writing

3. Review for Adequacy

    Compare with any Requirements
    Compare with Experience of Others

4. Revise as Needed

5. Get Approvals

    Consensus of Users
    Concurrence of Management
    Approval of Client/Agency

6. Implementation

7. On-going Review

LABORATORY QA PROGRAM DOCUMENT OUTLINE

1. Policy
   - Dedication to quality outputs
   - Commitment of adequate resources
   - Requirement for QA protocols for special purposes

2. Purpose
   - To satisfy concerns of analyst/management/clients
   - To inform on QA aspects

3. General Aspects
   - Requires planned work/experimentation
   - Requires GLP's/GMP's, o Requires Use of SOP's
   - Requires safe practices, o Requires safe disposal

4. Sampling
   - Stresses careful attention to samples and sampling
   - Stresses close identification of analytical data with sample considerations

5. Analytical Methodology
   - Stresses use of care in selection of methodology
   - Requires written procedures prior to their use in generating analytical data
   - Recommends written SOP's for recurring steps and methods
   - Stresses careful attention to calibration
   - Requires careful treatment of data

6. Laboratory Records
   - Emphasizes documentation of measurement details/data, maintenance records of equipment, filing of instruction manuals.

7. Control Charts
   - Affirms the importance of control charts for laboratory data operations

8. Quality Assessment
   - Affirms policy of validation by: multiple techniques, multiple operators, replicate measurements, use of SRM or other QC samples

9. Data Review/Reporting
   - Describes policy of data review and release
   - Describes general minimum content of all reports

10. Research Investigations
    - Recognizes the relation of quality data to quality research outputs
    - Affirms that research outpus require QA practices similar to analytical data outputs
    - Defines responsibilities

11. Implementation
    - Defines the respective responsibilities of individuals, management, and quality assurance coordinator for effective implementation of the QA program.

QUALITY ASSURANCE MANUAL

The Quality Assurance Manual is an Instruction kit that <u>you</u> have written for yourself <u>to guide</u> your laboratory operations in the <u>production</u> of QUALITY WORK

QUALITY ASSURANCE MANUAL

Content

    QA Program Document
    GLP's/GMP's
    SOP's
    Implementation Directives

Two Purposes

    Internal Guidance
    External Requirement

Should Be

    Brief
    Simple

Plusses

    Show Coordination or Need for Such
    Show Parallelisms or Lack Thereof

RELATING PERSONNEL TO QUALITY ASSURANCE

Quality Assurance Must Be a Way of Life
Quality Assurance Must Be Personal Objective
Quality Assurance Must Be Desired, not Accepted
Quality Assurance Must Be Looked Upon As a Necessary Part of

    Investigation

Quality Assurance Must Be Rewarding

    Not Punitive But,
    Recognition of Good Work
    Praise for Identifying Problems

Personnel Must Be Involved

    Consensus Development of GLP's/GMP's
    Concensus Development of SOP's

Educate - Skepticism Encouraged

PSYCHOLOGY OF QUALITY ASSURANCE

Analysts Should Develop Own Quality Assurance as far as Possible

Allotment of Time to Learn and Demonstrate Mastery of New Methods and on New Types of Samples

Cultivate Awareness and Appreciation of Statistical Concepts

Reward on the Basis of Quality

Role of Quality Circle

Keywords:

- o Knowledge
- o Ownership
- o Pride
- o Recognition

MODEL

Standard Operating Procedure
For

1. Introduction
    1.1 Purpose of Measurement/Test
        A brief description of why the measurement/test is needed and typical end-use(s) of the results/test report.

    1.2 Pre-Requisities

        1.2.1 Verification of Calibrations
            A brief statement of what calibrations must have been made previous to, or at the time of the measurement, and the checks required to verify that they have, indeed, been made and are valid at the time of use. This could include a check-list on the data sheet.

        1.2.2 Vertification of Equipment
            A statement of the checks, including trial measurements, that must be made, prior to a measurement sequence, to verify that the equipment is operating properly.

        1.2.3 Verification of Ability to Test
            A statement of the qualifying experience required before technical staff is permitted to make definitive measurement with, or apply the procedure to a specific test. The supervision that is required should be specified.

        Note: The data sheet may include a check-list to indicate that the various verifications have been made.

2. Methodology
    2.1 Standard Method
        Reference to a Standard Method as available. A copy should be appended.

    2.2 Laboratory Method
        When a standard method is not available, the method developed in the laboratory and previously tested for reliability should be included. This should follow the format of a standard method (such as ASTM format, for example). The following layout is considered to be minimal.

                Title
                Scope, Precision, Accuracy
                Summary
                Special Training
                Calibration/Standardization
                Procedure (with explanatory notes as needed)
                Critical Tolerances
                Calculations (include sample)
                Report
                Reference

MODEL SOP OUTLINE

Calibration

Purpose
Summary
Description of Item Calibrated
    Measurement Principle
    References to Manual

Calibration Interval
Equipment Needed
Standards Needed
    Source
    Preparation

Preliminary Operations
Procedure
    Reference to Standard Method When Applicable

Calculations
    Reduction of Data
    Uncertainty Limits

Report
    Format
    Labeling/Approval

Appendices
    Sample Report, References, etc.

MODEL SOP OUTLINE

Sampling Procedures

Introduction
    Critical Importance of the Sample
    Sampling Policy

General Instructions
    Sample Requirements
    Sampling Responsibilities
    Sample Preservation/Storage
    Sample Validation
    Sample Retention

Procedure
Records
    Labeling
    Logging
    Scheduling

Chain of Custody
Appendix
    References to Standard Practices
    Specific SOP's

PROTOCOLS FOR SPECIAL PURPOSES*

What They Are

    Protocols to Define What is to be Done in a Specific Case

How Prepared

    By Responsible Authority/Analyst
    May Concern Recurring Problems
    May be Tailored to a Specific Problem

        Approval by Management/Client Desirable in this Case

Content

    Specification of Principle Investigator (Study Director)
    Specification of Problem
    Specification of Model
    Specification of Sample
    Specification of Data Base
    Specification of Methodology
    Specification of any Deviations or Exceptions from GLP's/GMP's
    References to GLP's, GMP's, SOP's,. etc., as Appropriate
    Specification of Controls

        Control Charts to be Maintained
        Quality Assessment Procedures

    Release of Data

---

* Sometimes called Project QA Plans

QUALITY ASSURANCE COORDINATOR

Basic Function

The Quality Assurance Coordinator is responsible for the conduct of the quality assurance program and for taking or recommending corrective measures.

Responsibilities and Authority

1. Develops and carries out quality control programs, including statistical procedures and techniques, which will help meet desired quality standards at minimum cost.

2. Monitors quality assurance activities to determine conformance with policy and procedures and with sound practice; and makes appropriate recommendations for correction and improvement as may be necessary.

3. Seeks out and evaluates new ideas and current developments in the field of quality assurance and recommends means for their application wherever advisable.

4. Advises management in reviewing technology, methods, and equipment, with respect to quality assurance aspects.

5. Coordinates schedules for measurement system functional check calibrations, and other checking procedures.

6. Evaluates data quality and maintains records on related quality control charts, calibration records, and other pertinent information.

7. Coordinates and/or conducts quality-problem investigations.

Dr. John K. Taylor
September 18, 1984

PERSONAL QA PROFILE
Self Appraisal

This check list is designed for use of individuals to check their own QA profile by the self appraisal process. Each item should be considered and the score for the statement best describing the expertise/knowledge/performance should be entered in the box. Intermediate values may be chosen as appropriate. The level determinants are meant to be suggestive and are open to interpretation.

Tabulate the average score as indicated. An average score of 3.8 is acceptable but not laudatory. A score of 2.5 or lower is considered to be unacceptable and indicates major QA deficiencies. Intermediate scores indicate the need for immediate remedial actions.

No matter what average score is obtained, individuals should examine the scores for individual items to identify QA areas that need improvement. Any item rated at 3 or below should be so considered.

PERSONAL QA PROFILE
Self Appraisal

1. Knowledge of Field /___/
   - State-of-the-art knowledge of my field — 5
   - Good practical knowledge of field — 3
   - Some gaps/limited understanding — 1

2. General Understanding of Methodology Used /___/
   - Excellent comprehensive knowledge of methodology, including basic theory — 5
   - Make point to understand methodology, before use — 3
   - Practical understanding but some gaps in basic comprehension — 1

3. Mastery of Specific Technology /___/
   - State-of-the-art accuracy and precision always attained — 5
   - Average accuracy and precision attained — 3
   - Accuracy and precision needs improvement — 1

4. Use of Written SOP's /___/
   - Use SOP's and/or develop written procedures for all methods used — 5
   - Use SOP's/written methods for all critical analyses — 3
   - Seldom or little use of written methods — 1

5. Pre-check of Methodology Prior to Use /___/
   - Extensive checks of new methods/pre-analysis check of all others, before use — 5
   - Pre-checks confined to new methodology — 3
   - Little or no pre-checking — 1

6. Adherence to GLP's/GMP's /___/
   - GLP's/GMP's developed and used regularly — 5
   - GLP's/GMP's for most critical operations — 3
   - GLP's/GMP's little/not used — 1

7. Laboratory Notebooks /‾‾/
    Neat, indexed lab notebooks maintained, readily understandable
        to others   5
    Some deficiencies in notebooks but generally acceptable   3
    Notebooks need considerable improvement   1

8. Knowledge of Statistics /‾‾/
    Full working knowledge and extensive use of statistics in all decision
        processes   5
    Good understanding and occasional use of statistics   3
    Limited knowledge of statistics   1

9. Control Chart Usage /‾‾/
    Good understanding, extensive use of control charts   5
    Occasional use of control charts   3
    Limited/little use of control charts   1

10. Participation in Technical Activities /‾‾/
    Active/leadership role in technical organizations   5
    Passive Role   3
    Little participation   1

11. Training Courses /‾‾/
    Training course(s) taken during past 18 months   5
    Training course(s) taken during past 3 years   3
    No recent training taken   1

12. Technical Books/Informal Training /‾‾/
    Informal training/tech books read during past year   5
    Some informal training during past 2 years   3
    No recent informal training   1

13. Experimental Planning /‾‾/
    Experimental planning understood/used extensively in all
        major activities   5
    Work generally well planned before starting   3
    Planning needs considerable improvement   1

14. Use of Randomization /‾‾/
    Understand/always use randomization in work plans/execution of work   5
    Some use of randomization in work plans   3
    Limited/little use of randomization concepts   1

15. Housekeeping Practices /‾‾/
    Work space always tidy consistent with activities in progress   5
    No major problems in housekeeping   3
    Housekeeping not a strong point   1

Average Score   /‾‾/

Dr. John K. Taylor
September 17, 1984

Laboratory QA Profile
Self Appraisal

This check sheet is designed for use by laboratory management to appraise the QA program of the laboratory. Each item should be considered individually and the appropriate score entered in the box. Intermediate values may be chosen. The level determinants are meant to be suggestive and are open to interpretation.

An average score of 3.8 is acceptable but not laudatory. A score of 2.5 or lower is considered to be unacceptable and indicates that serious risk exists in laboratory operations. Intermediate scores will require a review of the QA program to identify and rectify major deficiencies.

Even in the case of an acceptable average score, a low score for any item (≤ 3) should be considered for possible corrective actions.

LABORATORY QA PROFILE
Self Appraisal

1. Laboratory QA Program /____/
   - Written plan adopted/implemented/in use — 5
   - Definite but informal program — 3
   - Informal/variable program — 1

2. Use of Written (Before Use) Methodology /____/
   - Exclusively — 5
   - Majority of time/for all critical data — 3
   - Few or none used — 1

3. Control Chart Use /____/
   - Maintained for all critical operations — 5
   - Variable but significant use in organization — 3
   - Little or no use — 1

4. Uncertainty Limits for Data /____/
   - Limits for all data outputs/policy/enforced — 5
   - Most of the time/at least where critical — 3
   - Minority of cases — 1

5. Reports/Proposals /____/
   - Pre- and post-screened for QA aspects — 5
   - Those deemed critical are screened — 3
   - Variable/seldom done — 1

6. Facilities Maintenance /____/
   - Excellent (showplace condition) — 5
   - Good (passes muster) — 3
   - Poor (reservations, no-no areas) — 1

7. Equipment Maintenance /____/
   - Regular maintenance with records kept, control charts as appropriate — 5
   - Good maintenance, documentation of such has some deficiencies — 3
   - Irregular maintenance practices — 1

8. Records /____/
   - Laboratory records judged to be excellent by any standards — 5
   - Some reservations; could be difficulties in spots — 3
   - Variable, need considerable improvement — 1

9. Training New Employees   /___/
       Formal QA indoctrination   5
       Informal QA indoctrination   3
       Assumed not needed/not done   1

10. Personnel/Staff QA Consciousness*   /___/
       Staff average for personnel QA audit ≥ 4   5
       Staff average for personnel QA audit 3 to 4   3
       Staff average for personnel QA audit ≤2.5   1
          *Alternatively, average staff QA audit
          values may be inserted in /___/

11. Professional Interations   /___/
       Majority and all key staff active in some professional organization 5
       Reasonable level of activity   3
       Little or don't know   1

12. Management and Statistics   /___/
       High level of knowledge and ability to use at supervisory level and above 5
       General awareness and reasonable usage   3
       Variable comprehension/use   1

13. Internal QA Audits   /___/
       Regular program/feedback/corrective actions   5
       Occasional audits   3
       Few or none   1

14. External QA Audits   /___/
       Regular external appraisal of QA policy/practices   5
       QA appraisal definite part of other reviews   3
       None   1

15. Overall Opinion of QA Status   /___/
       No known weaknesses that are not subject of corrective actions   5
       Known QA weaknesses but less than vigourous action to correct them 3
       No or little basis for judgement   1

Average Score   /___/

APPENDIX A

APPENDIX A

Quality Assurance of Chemical Measurements

John K. Taylor
Center for Analytical Chemistry
National Bureau of Standards
Washington, D.C. 20234

John K. Taylor
Center for Analytical Chemistry
National Bureau of Standards
Washington, D.C. 20234

Quality Assurance of

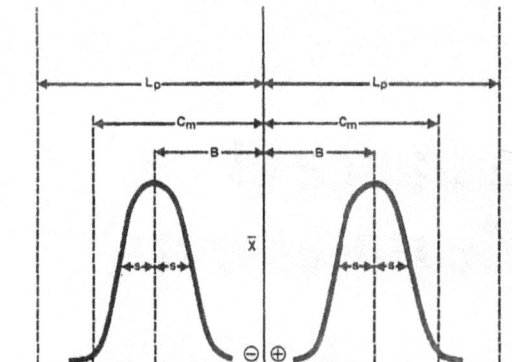

Figure 1. Measurement tolerances and errors

The objective of quality assurance programs for analytical measurements is to reduce measurement errors to tolerable limits and to provide a means of ensuring that the measurements generated have a high probability of being of acceptable quality. Two concepts are involved. *Quality control* is the mechanism established to control errors, while *quality assessment* is the mechanism to verify that the system is operating within acceptable limits. General handbooks that discuss quality assurance in more detail are given in References 1–3.

Quality is a subjective term. What is high quality in one situation may be low or unacceptable quality in another case. Clearly the tolerable limits of error must be established for each. Along with this there must be a clear understanding of the measurement process and its capability to provide the results desired.

The tolerance limits for the property to be measured are the first conditions to be determined. These are based upon the considered judgment of the end user of the data and represent the best estimate of the limits within which the measured property must be known, to be useful for its intended purpose. The limits must be realistic and defined on the basis of cost–benefit considerations. It is better to err on the side of too-narrow limits. Yet, measurement costs normally increase as tolerances are decreased, so that the number of measurements possible for a fixed budget may be inadequate when coupled with material-variability considerations.

Once one has determined the tolerance limits for the measured property, the permissible tolerances in measurement error may be established. The basis for this is shown in Figure 1. The tolerance limits for the measured property are indicated by L_p. Uncertainties in the measurement, based on the experience and judgment of the analyst, are indicated by C_m. These include estimates of the bounds for the biases (systematic errors), B, and the random errors as indicated by s, the estimate of the standard deviation. Obviously, C_m must be less than L_p if the data are to be useful. The confidence limits for $\bar{x}$, the mean of n replicate measurements, are:

$$C_m = \pm \left[B + \frac{ts}{\sqrt{n}} \right]$$

in which t is the so-called student factor. While the effect of random error is minimized by replication of measurements, there are practical limitations, and any measurement process that requires a large number of replicates has a serious disadvantage.

Well-designed and well-implemented quality assurance programs provide the means to operate a measurement system in a state of statistical control, thereby providing the basis for establishing reliable confidence limits for the data output.

Until a measurement operation ... has attained a state of statistical control, it cannot be regarded in any logical sense as measuring anything at all.

C. E. Eisenhart

The Analytical System

Analytical measurements are made because it is believed that compositional information is needed for some end use in problem solving. Explicitly or implicitly, a measurement system such as that depicted in Figure 2 is involved. One must have full understanding of the measurement system for each specific situation in order to generate quality data.

The conceptualization of the problem, including the data requirements and their application, constitutes the model. The plan, based on the model, includes details of sampling, measurement, calibration, and quality assurance. Various constraints such as time, resources, and the availability of samples may necessitate compromises in the plan. Adequate planning will require the collaboration of the analyst, the statistician, and the end user of the data in all but the most routine

Report

Chemical Measurements

cases. In complex situations, planning may be an iterative process in which the actual data output may require reconsideration of the model and revision of the plan.

Sampling has been discussed in a recent paper (4). Obviously, the sample is one of the critical elements of the measurement process. Closely related is the measurement methodology to be used. The method used must be adequate for the intended purpose and it must be properly utilized. The necessary characteristics of a suitable method include: adequate sensitivity, selectivity, accuracy, and precision. It is desirable that it also have the following characteristics: large dynamic measurement range; ease of operation; multiconstituent applicability; low cost; ruggedness; portability. To judge its suitability, the following information must be known about it: type of sample; forms determined; range of applicability; limit of detection; biases; interferences; calibration requirements; operational skills required; precision; and accuracy. Obviously all of the above characteristics must match the measurement requirements. In case of doubt, trial measurements must be made to demonstrate applicability to a given problem. A cost-benefit analysis may be needed to determine which of several candidate methods is to be selected. A method, once adopted, must be used in a reliable and consistent manner, in order to provide reproducible data. This is best accomplished by following detailed written procedures called Standard Operating Procedures (SOPs) in quality assurance terminology. Standard methods developed by voluntary standardization organizations are often good candidates for SOPs, when they are available.

Two kinds of calibrations are required in most cases. Physical calibrations may be needed for the measurement equipment itself and for ancillary measurements such as time, temperature, volume, and mass. The measurement apparatus may include built-in or auxiliary tests such as voltage checks, which may need periodic verification of their stability if not of their absolute values. But especially, most analytical equipment requires some kind of chemical calibration, often called standardization, to establish the analytical function (i.e., the relation of instrument response to chemical quantification). Obviously, the analyst must thoroughly understand each of the calibrations required for a particular measurement. This includes a knowledge of the standards needed and their relation to the measurement process, the frequency of calibration, the effect on a measurement system due to lack of calibration, and even the shock to the system resulting from recalibration.

Quality Control

Quality control encompasses all of the techniques used to encourage reproducibility of the output of the measurement system. It consists of the use of a series of protocols developed in advance and based on an intimate understanding of the measurement process and the definite requirements of the specific measurement situation. Protocols, i.e., procedures that must be rigorously followed, should be established for sampling, measurement, calibration, and data handling. Some of these, or at least selected portions, may be applicable to most or all of the measurements of a particular laboratory and become the basis of a good laboratory practices manual (GLPM).

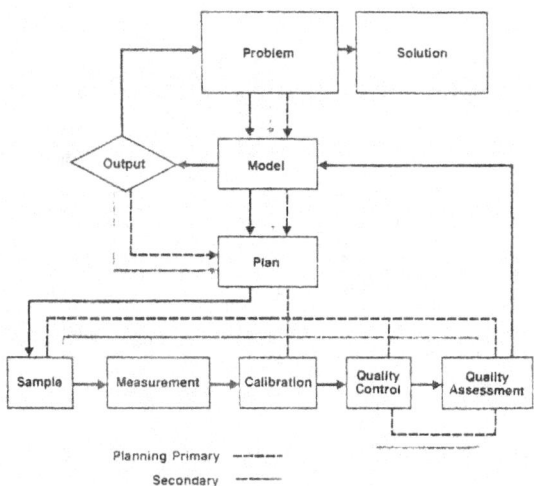

Figure 2. Analytical measurement system

Figure 3. Quality control by inspection

In fact, the GLPM should cover the generalities, if not the specifics, of all measurement practices of the laboratory. The protocols for a specific measurement process include the GLPs together with any requirements of the specific situation.

The GLPM and protocols should be developed collaboratively by all of those involved in the measurements, and this development process may be the most important aspect of their function. It encourages a keen consideration of the measurement process and creates an awareness of potential problems that GLPs attempt to avert.

Protocols are of little use unless they are followed rigorously, and the attitudes of laboratory personnel are certainly key factors in this regard. Analysts must aspire to produce high quality data and must be their own most severe critics. Notwithstanding, good quality control systems should include provisions for inspection, both periodically and aperiodically (unannounced) to ascertain how well they are functioning. Large laboratories may have a quality control officer or group, independent of the laboratory management, that oversees the operation of the quality control system.

Quality Control by Inspection

An informal kind of quality control involves the frequent if not constant inspection of certain aspects of the measurement system for real or apparent problems (5). The essential features of such a system are depicted in Figure 3. Based on an intimate knowledge of the measurement process, samples may be casually inspected for their adequacy. The rejection and possible replacement of obviously unsuitable ones can eliminate not only extra work but also erroneous data that might be difficult to identify later. Difficulties in the actual measurement may often be identified as they occur and remedial measures, including remeasurement, may be taken either to save data that might otherwise be lost or at least to provide valid reasons for any rejections. Likewise, data inspection can identify problems and initiate remedial actions, including new measurements, while it is still possible to do so.

Control Charts

The performance of a measurement system can be demonstrated by the measurement of homogeneous and stable control samples in a planned repetitive process. The data so generated may be plotted as a control chart in a manner to indicate whether the measurement system is in a state of statistical control. Either the result of a single measurement on the control sample, the difference between duplicate measurements, or both may be plotted sequentially. The first mode may be an indicator of both precision and bias, while the second monitors precision only.

To effectively use such a chart, the standard deviation of a single measurement of the control sample must be known. This may be obtained by a series of measurements of the control sample, or it may be obtained from the experience of the laboratory on measuring similar samples. Control limits, i.e., the extreme values believed to be credible, are computed from the standard deviation. For example, the 2σ limit represents those within which the values are expected to lie 95% of the time. The 3σ limit represents the 99.7% confidence level. Departures from the former are warnings of possible trouble, while exceeding the latter usually means corrective action is needed. In the event that the standard deviation cannot be estimated with sufficient confidence initially, the control chart may be drawn using the best estimate, and the limits may be modified on the basis of increasing measurement experience.

The development of a control chart must include the rationale for its use. There must be a definite relation between the control measurements and the process they are designed to control. While the control chart only signifies the degree of replication of measurements of the control sample, its purpose is to provide confidence in the measurement process. To do this, the control measurements must simulate the measurements normally made. In chemical measurements, this means simulation of matrix, simulation of concentration levels, and simulation of sampling. The latter objective may be difficult if not impossible to achieve. It must be further emphasized that the control measurements should be random members of the measurement routine, or at least they should not occupy biased positions in any measurement sequence.

To the extent that control samples are representative of the test samples, and to the extent that measurements of them are representative of the measurement process, the existence of statistical control for these samples can imply such control of the measurement process and likewise of the results obtained for the test samples.

No specific statements can be made about the frequency of use of control samples. Until a measurement process is well understood, control samples may need to be measured frequently. As it is demonstrated to be in control, the need may become less and the incentive to do "extra" work may diminish. Along with the decision on how much effort should be devoted to quality control the risks and consequences of undetected loss of control must be weighed. Many laboratories consider that the 5–15% extra effort ordinarily required for all aspects of quality control is a small price to pay for the quality assurance it provides. When measurements are made on a frequently recurring schedule, internal controls, such as duplicate measurements of test samples, can provide evidence of reproducibility so that control samples may be used largely to identify systematic errors, drifts, or other types of problems.

When laboratories are engaged in a variety of measurements, the use of representative control samples may be difficult if not impossible. In such cases, often only the measurement methodology can be tested, and evaluation of the quality of the measurement output requires considerable judgment. In such cases, the experience of the lab becomes a key factor.

In some complex measurement systems, certain steps or subsystems are more critical than others, and hence it may be more important to develop control charts for them than for the entire system. The control of such steps may indeed prevent propagation of error into the end result. An example is the sampling step, which may be very critical with respect to the end result. In such a case, the records of periodic inspections may be adaptable to the control chart technique of quality control.

Quality Assessment

Procedures used to evaluate the effectiveness of the quality control system may be classified according to whether the evidence arises from internal or external sources. Internal procedures, useful largely for estimating precision, include the use of internal reference samples and control charts to monitor the overall performance of the measurement system as described in an earlier section. Replicate measurements on replicate or split samples can provide valuable insight into the reproducibility of both the measurement and sampling processes. Comparison of the results obtained as a consequence of interchange of analysts, equipment, or combinations of these can attest to operational stability as well as identify malfunctions. Measurements made on split samples using a completely independent method can lend confidence to the method normally in use or indicate the presence of measurement bias.

External quality assessment is always needed since it can detect problems of bias that are difficult to identify by internal procedures. Participation in collaborative tests, exchange of samples with other laboratories, and the use of certified reference materials are time-honored assessment devices. NBS Standard Reference Materials (SRMs) (6) are especially useful for quality assessment in cases where they are available and applicable. The information that can be obtained or inferred by their use is described in a later section. Operators of monitoring networks may provide proficiency testing or audit samples to assess laboratory performance. Ordinary practices should be used here, so that normal rather than optimum performance is measured.

A laboratory should diligently use the information obtained in the quality assessment process. Adverse data should not be treated in a defensive manner but the reason for it should be investigated objectively and thoroughly. When laboratory records are reliably and faithfully kept, the task of identifying causes of problems is made easier. This is an important reason for developing data handling protocols and ensuring that all protocols are strictly followed.

Systematic Errors

Systematic errors or biases are of two kinds—concentration-level independent (constant), and concentration-level related. The former are sometimes called additive while the latter are called multiplicative. Both kinds may be present simultaneously in a given measurement system. An example of the first kind is the reagent blank often present in measurements involving chemical processing steps. The second kind can result from, for example, use of an inaccurately certified calibrant.

Systematic errors may arise from such sources as faulty calibrations, the use of erroneous physical constants, incorrect computational procedures, improper units for measurement or reporting data, and matrix effects on the measurement. Some of these can be eliminated or minimized by applying corrections or by modification of the measurement technique. Others may be related to fundamental aspects of the measurement process. The most insidious sources of error are those unknown or unsuspected of being present.

One of the most important sources of error in modern instrumental measurements concerns uncertainties in the calibrants used to define the analytical function of the instrument. The measurement step essentially consists of the comparison of an unknown with a known (calibrant) so that any error in the latter results in a proportional error in the former. The need to use calibrants of the highest reliability is obvious.

The measurement protocol should include a detailed analysis of the sources of error and correction for them to the extent possible. The uncertainties, B, referred to earlier, represent the uncertainties in the corrections for the systematic errors. In making such an estimate, the 95% confidence limits should be assigned to the extent possible. The magnitudes of these uncertainties can be estimated from those assigned by others in the case of such factors as calibration standards and physical constants. Other constant sources of error may be more subtle both to identify and to evaluate, and the judgment and even intuition of the experimenter may be the only sources of information.

The effectiveness of elimination of, or correction for, systematic errors is best evaluated from external quality assessment procedures. Differences found between known and measured values of test samples, such as SRMs, need to be reconciled with the laboratory's own estimates of bounds for its random and systematic errors. When the random error is well established, as by the quality control process, significant discrepancies can be attributed to unsuspected or incorrectly estimated systematic errors.

The Use of SRMs for Quality Assessment

An SRM is a material for which the properties and composition are certified by the National Bureau of Standards (6, 7). To the extent that its compositional properties simulate

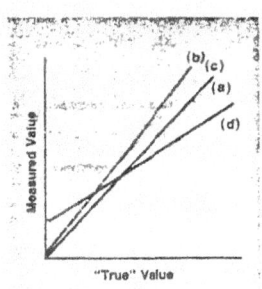

Figure 4. Typical analytical systematic errors (bias). (a) = unbiased; (b) = measurement-level related; (c) = constant error; and (d) = combination of b and c

those of the sample ordinarily measured, its "correct" measurement can imply "correct" measurement of the usual samples. Such a conclusion requires that the protocol of measurement was the same in each case. Hence it is necessary that no special care be exercised in measuring the SRM, other than that ordinarily used.

Analysis of SRMs has been recommended as a means of providing "traceability" to national measurement standards. However, a word of caution is appropriate on this point. Measurement processes are seldom identical, so that traceability is most often based on inference. Also, the fact that an acceptable result is or is not obtained for an SRM provides no unique explanation for such a result.

The use of an SRM should never be attempted until the analytical system has been demonstrated to be in a state of statistical control. An SRM is not needed for such a purpose and such use is discouraged. Ordinarily, the SRM will be available in limited amount so that the statistics of the measurement process should be demonstrated by measurements on other materials. Only under such a situation can the results of an SRM measurement be considered as representative of the measurement system.

A consideration of the nature of analytical errors, shown in Figure 4, will clarify why the measurement of a single SRM may not be fully informative. It will be noted that errors may be constant, measurement-level related, or a combination of these, and a single right or wrong result will not indicate on which of several possible curves it might lie. Measurement of a series of SRMs may clarify the nature of the measurement process and this should be done whenever possible. An intimate understanding of the operation of a particular measurement system may also make it possible to eliminate some of the possible sources of error and to better interpret the data from measurement of SRMs.

A-5

Record Keeping

Adequate record keeping in an easily retrievable manner is an essential part of the quality assurance program. Records needed include the description of test samples, experimental procedures, and data on calibration and testing. Quality control charts should be diligently prepared and stored. A chain of custody of test materials should be operative and such materials should be retained and safeguarded until there is no doubt about their future use or need.

Data Control

The evaluation, review, and release of analytical data is an important part of the quality assurance process. No data should be released for external use until it has been carefully evaluated. Guidelines for data evaluation, applicable to almost every analytical situation, have been developed by the ACS Committee on Environmental Improvement (8). A prerequisite for release of any data should be the assignment of uncertainty limits, which requires the operation of some kind of a quality assurance program. Formal release should be made by a professional analytical chemist who certifies that the work was done with reasonable care and that assigned limits of uncertainty are applicable.

Laboratory Accreditation

Laboratory accreditation is one form of quality assurance for the data output of certified laboratories. Accreditation is based on criteria that are considered essential to generate valid data and is a formal recognition that the laboratory is competent to carry out a specific test or specific type of test (9, 10). The certification is as meaningful as the care exercised in developing certification criteria and evaluating laboratory compliance. Generic criteria developed by national and international standardization organizations have been influential in this respect (11). These criteria are well conceived and provide general guidance for the sound operation of analytical laboratories, whether or not certification is involved.

Implementation

Detailed quality assurance plans are ineffective unless there is commitment to quality by all concerned. This commitment must be total, from management to technical staff. The former must provide the resources, training, facilities, equipment, and encouragement required to do quality work. The latter must have the technical ability and motivation to produce quality data. Some may argue that if there is such commitment, there is no need for a formal quality assurance program. However, the experience of many laboratories has demonstrated that a formal quality assurance program provides constant guidance for the attainment of the quality goals desired.

References

(1) "Quality Assurance Handbook for Air Pollution Measurement Systems: Principles"; Vol. 1, E.P.A. Publication No. 600/9-76-005.
(2) "Handbook for Analytical Quality Control in Water and Wastewater Laboratories"; EPA Publication No. 600/4-79-019.
(3) Juran, J. M. "Quality Control Handbook," 3rd ed.; McGraw-Hill: New York, 1974.
(4) Kratochvil, B. G.; Taylor, J. K. Anal. Chem. 1981, 53, 924 A.
(5) Brewers, J. M., et al. "Data Are for Looking At, or Quality Control by Interpretation." In "Water Quality Parameters"; American Society for Testing and Materials: Philadelphia, 1975, ASTM STP573.
(6) "Catalogue of NBS Standard Reference Materials"; National Bureau of Standards: Washington, NBS Special Publication 260.
(7) Cali, J. P., et al. "The Role of Standard Reference Materials in Measurement Systems"; National Bureau of Standards: Washington, January 1975, NBS Monograph 148.
(8) ACS Subcommittee on Environmental Analytical Chemistry. "Guidelines for Data Acquisition and Data Quality Evaluation in Analytical Chemistry"; Anal. Chem. 1980, 52, 2242.
(9) "Quality Control System-Requirements for a Testing and Inspections Laboratory"; American Council of Independent Laboratories: Washington.
(10) "Testing Laboratory Performance: Evaluation and Accreditation"; Berman, G. A., Ed.; National Bureau of Standards: Washington. NBS Special Publication 591.
(11) "Standard Recommended Practice for Generic Criteria for Use in Evaluation of Testing and/or Inspection Agencies"; American Society for Testing and Materials: Philadelphia, ASTM Publication No. E-548.

John K. Taylor, coordinator for quality assurance and voluntary standardization activities at the National Bureau of Standards Center for Analytical Chemistry, received his BS from George Washington University, and his MS and PhD degrees from the University of Maryland. His research interests include electrochemical analysis, refractometry, isotope separations, standard reference materials, and the application of physical methods to chemical analysis.

APPENDIX B

APPENDIX B

Sampling for Chemical Analysis

A major consideration in the reliability of any analytical measurement is that of sample quality. Too little attention is directed to this matter. The analyst often can only report results obtained on the particular test specimen at the moment of analysis, which may not provide the information desired or needed. This may be because of uncertainties in the sampling process, or in sample storage, preservation, or pretreatment prior to analysis. The sampling plan itself is often so poorly considered as to make relation of the analytical results to the population from which the sample was drawn uncertain, or even impossible to interpret.

All of the above aspects of sampling merit full consideration and should be addressed in every analytical determination. Because the scope is so broad, we will limit the present discussion to a small segment of the total problem, that of sampling bulk materials. For such materials the major steps in sampling are:
* identification of the population from which the sample is to be obtained,
* selection and withdrawal of valid gross samples of this population, and
* reduction of each gross sample to a laboratory sample suitable for the analytical techniques to be used.

The analysis of bulk materials is one of the major areas of analytical activity. Included are such problems as the analysis of minerals, foodstuffs, environmentally important substances, and many industrial products. We shall discuss the major considerations in designing sampling programs for such materials. While our discussion is specifically directed toward solid materials, extension to other materials will often be obvious.

A brief list of definitions commonly used in bulk sampling is provided in the glossary.

Preliminary Considerations in Sampling

Poor analytical results may be caused in many ways—contaminated reagents, biased methods, operator errors in procedure or data handling, and so on. Most of these sources of error can be controlled by proper use of blanks, standards, and reference samples. The problem of an invalid sample, however, is special; neither control nor blank will avail. Accordingly, sampling uncertainty is often treated separately from other uncertainties in an analysis. For random errors the overall standard deviation, s_o, is related to the standard deviation for the sampling operation, s_s, and to that for the remaining analytical operations, s_a, by the expression: $s_o^2 = s_a^2 + s_s^2$. Whenever possible, measurements should be conducted in such a way that the components of variance arising from sample variability and measurement variability can be separately evaluated. If the measurement process is demonstrated to be in a state of statistical control so that s_a is already known, s_s can be evaluated from s_o found by analysis of the samples. Otherwise, an appropriate series of replicate measurements or replicate samples can be devised to permit evaluation of both standard deviations.

Youden has pointed out that once the analytical uncertainty is reduced to a third or less of the sampling uncertainty, further reduction in the analytical uncertainty is of little importance (1). Therefore, if the sampling uncertainty is large and cannot be reduced, a rapid, approximate analytical method may be sufficient, and further refinements in the measurement step may be of negligible aid in improving the overall results. In fact, in such cases a rapid method of low precision that permits more samples to be examined may be the best route to reducing the uncertainty in the average value of the bulk material under test.

An excellent example of the importance of sampling is given in the deter-

Published in Analytical Chemistry, July 1981, pp. 924A–938A, by the American Chemical Society

Report

Byron Kratochvil
Department of Chemistry
University of Alberta
Edmonton, Alberta, Canada T6G 2G2

John K. Taylor
National Bureau of Standards
Washington, D.C. 20234

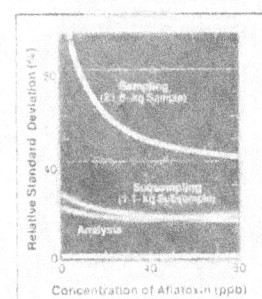

Figure 1. Relative standard deviation associated with the sampling and analysis operations in testing peanuts for aflatoxins (after T. B. Whittaker, *Pure and Appl. Chem.*, **49**, 1709 (1977))

mination of aflatoxins in peanuts (2). The aflatoxins are highly toxic compounds produced by molds that grow best under warm, moist conditions such conditions may be localized in a warehouse, resulting in a patchy distribution of highly contaminated kernels. One badly infected peanut can contaminate a relatively large lot with unacceptable levels (above about 25 ppb for human consumption) of aflatoxins after grinding and mixing. The standard deviations of the three operations of sampling, subsampling, and analysis are shown in Figure 1. The analytical procedure consists of solvent extraction followed by thin-layer chromatography and measurement of the fluorescence of the aflatoxin spots. Clearly, sampling is the major source of the analytical uncertainty.

Types of Samples

Random Samples. In common with the statistician, the analytical chemist ordinarily wishes to generalize from a small body of data to a larger body of data. While the specimen sample actually examined is sometimes the only matter of interest, the characteristics of the population of specimens are frequently desired. Obviously, the samples under examination must not be biased, or any inferences made from them will likewise be biased.

Statisticians carefully define several terms that are applied to statistical inference. The *target population* denotes the population to which we would like our conclusions to be applicable, while the *parent population* designates that from which samples were actually drawn. In practice these two populations are rarely identical, although the difference may be small. This difference may be minimized when the selection of portions for examination is done by a random process. In such a process each part of the population has an equal chance of being selected. Thus random samples are those obtained by a random sampling process, and form a foundation from which general inferences based on mathematical probability can be made.

Random sampling is difficult. A sample selected haphazardly is not a random sample. On the other hand, samples selected by a defined protocol are likely to reflect the biases of the protocol. Even under the most favorable circumstances, unconscious selection and biases can occur. Also, it can be difficult to convince intransigent individuals that the risk of obtaining samples that an apparently unsystematic selection pattern must be followed rigorously if it is to be used.

Whenever possible, the use of a table of random numbers is recommended as an aid to sample selection. The bulk material is divided into a number of real or imaginary segments. For example, a body of water can be conceptually subdivided into cells, both horizontally and vertically, and the cells to be sampled selected randomly. To do this each segment is assigned a number, and selection of segments from which sample increments are to be taken is made by starting in an arbitrary place in a random number table and choosing numbers according to a predecided pattern. For example, one could choose adjacent, alternate, or nth entries and sample those segments whose numbers occur until all of the samples decided upon have been obtained.

The results obtained for these and other random samples can be analyzed by some model or plan to identify whether systematic relations exist. This is important because of the possible introduction of apparent correlations due to systematic trends or biases in the measurement process. Accordingly, measurement plans should always be designed to identify and minimize such problems.

Despite the disadvantages, sampling at evenly-paced intervals over the bulk is still often used in place of random sampling owing to its simplicity

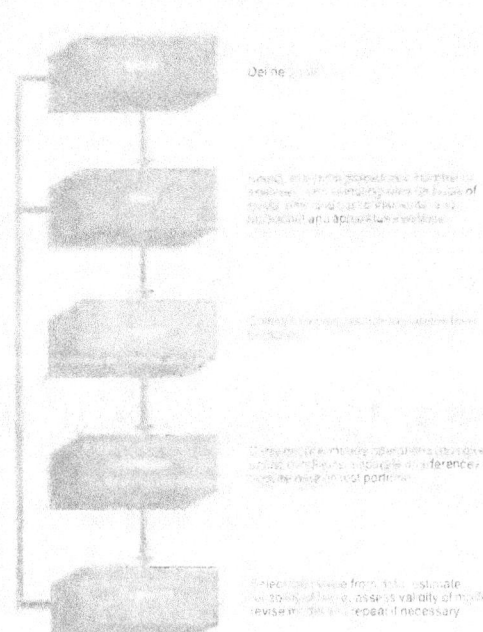

Figure 2. The role of sampling in the overall analytical process

Because of the limited information provided by a composite sample, full consideration should be given to the consequences before deciding between this approach and the analysis of individual samples.

Subsampling. Usually, the sample received by the analytical laboratory will be larger than that required for a single measurement, so some subsampling (see glossary) will be required. Often, test portions (see glossary) must be taken for replicate measurements or for measurement of different constituents by several techniques. Obviously, such test portions must be sufficiently alike that the results are compatible. Frequently it is necessary to reduce particle size, mix, or otherwise process the laboratory sample (see glossary) before withdrawing portions (subsamples) for analysis. The effort necessary at this stage depends on the degree of homogeneity of the original sample. In general, the subsampling standard deviation should not exceed one-third of the sampling standard deviation. Although this may sound appreciable, it is wasteful of time and effort to decrease it below this level. But this does not mean care is unnecessary in subsampling. If a sample is already homogeneous, care may be needed to avoid introducing segregation during subsampling. Even though analysts may not be involved with sample collection, they should have sufficient knowledge of sampling theory to subsample properly. They should also be provided with any available information on the homogeneity of the samples received so that they can subsample adequately and efficiently.

Model of the Sampling Operation

Before sampling is begun, a model of the overall operation should be established (Figure 2). The model should consider the population to be studied, the substance(s) to be measured, the extent to which speciation is to be determined, the precision required, and the extent to which the distribution of the substance within the population is to be obtained.

The model should identify all assumptions made about the population under study. Once the model is complete, a sampling plan can be established.

The Sampling Plan

The plan should include the size, number, and location of the sample increments and, if applicable, the extent of compositing to be done. Procedures for reduction of the gross sample (see glossary) to a laboratory sample, and to the test portions, should be specified. All of this should be written as a

Table I. Confidence Intervals and Statistical Tolerance Limits[a]

n	t[b]	$\frac{ts}{\sqrt{n}}$	K[c]	Ks
2	12.70	±18	37.67	±75
4	3.18	±3.2	6.37	±12.9
8	2.36	±1.7	3.73	±7.4
16	2.13	±1.1	2.90	±5.8
32	2.04	±0.7	2.50	±5.0
100	1.98	±0.4	2.23	±4.4
∞	1.96	0	1.96	±4.0

[a] Calculated for $s = 2$, based on measurement of n samples
[b] 95% confidence limits for the mean of n samples
[c] Based on a 95% confidence that the interval will contain 95% of the samples

detailed protocol before work is begun. The protocol should include procedures for all steps, from sampling through sample treatment, measurement, and data evaluation; it should be revised as necessary during execution as new information is obtained. The guidelines for data acquisition and quality evaluation in environmental chemistry set out by the ACS Subcommittee on Environmental Analytical Chemistry are sufficiently general to be recommended reading for workers in all fields (4).

The sampling protocol should include details of when, where, and how the sample increments are to be taken. On-site criteria for collection of a valid sample should be established beforehand. Frequently, decisions must be made at the time of sampling as to components likely to appear in the sample that may be considered foreign, that is, not part of the population. For example, a portion of a dredged sediment in which the mercury content is to be determined might contain cans, discarded shoes, rocks or other extraneous material. For the information sought these items might be considered foreign and therefore legitimately rejected. Decisions as to rejection become less clear with smaller items. Should smaller stones be rejected? How small? And what about bits of metal, glass, leather, and so on? Criteria for such decisions should be made logically and systematically, if possible before sampling is initiated.

The type of container, cleaning procedure, and protection from contamination before and after sampling must be specified. The question of sample preservation, including possible addition of preservatives and refrigeration, should be addressed. Some sampling plans call for field blanks and/or field-spiked samples. The critical nature of the latter and the difficulties possible under field conditions require the utmost care in planning and execution of the sampling operation if the results are to be meaningful.

Whenever possible, the analyst should perform or directly supervise the sampling operation. If this is not feasible, a written protocol should be provided and the analyst should ensure that those collecting the samples are well-trained in the procedures and in use of the sampling equipment, so that bias and contamination are minimized. No less important is careful labeling and recording of samples. A chain of custody should be established such that the integrity of the samples from source to measurement is ensured. Often auxiliary data must be recorded at the time the sample is taken: temperature, position of the collecting probe in the sample stream, flow velocity of the stream, and so on. Omission or loss of such information may greatly decrease the value of a sample, or even render it worthless.

Sampling Bulk Materials. Once the substances to be determined, together with the precision desired, have been specified, the sampling plan can be designed. In designing the plan, one must consider:
• How many samples should be taken?
• How large should each be?
• From where in the bulk material (population) should they be taken?
• Should individual samples be analyzed, or should a composite be prepared?

These questions cannot be answered accurately without some knowledge of the relative homogeneity of the system. Gross samples should be unbiased with respect to the different sizes and types of particles present in the bulk material. The size of the gross sample is often a compromise based on the heterogeneity of the bulk material on the one hand, and the cost of the sampling operation on the other.

When the properties of a material

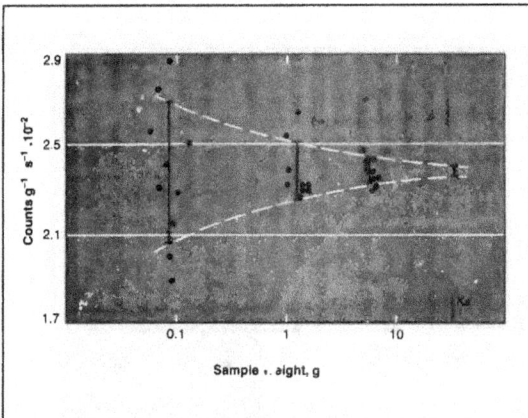

Figure 3. Sampling diagram of sodium-24 in human liver homogenate (from Reference 7)

to be sampled are unknown, a good approach is to collect a small number of samples, using experience and intuition as a guide to making them as representative of the population as possible, and analyze for the component of interest. From these preliminary analyses, the standard deviation s_s of the individual samples can be calculated, and confidence limits for the average composition can be established using the relation

$$\mu = \bar{x} \pm ts_s/\sqrt{n} \qquad (1)$$

where μ is the true mean value of the population, $\bar{x}$ is the average of the analytical measurements, and t is obtained from statistical tables for n measurements (often given as $n - 1$ degrees of freedom) at the desired level of confidence, usually 95%. Table I lists some t values; more extensive tables are provided in books on quantitative analysis and statistics (5).

On the basis of this preliminary information, a more refined sampling plan can be devised, as described in the following sections. After one or two cycles the parameters should be known with sufficient confidence that the optimum size and number of the samples can be estimated with a high level of confidence. The savings in sampling and analytical time and costs by optimizing the sampling program can be considerable.

Minimum Size of Individual Increments. Several methods have been developed for estimation of the amount of sample that should be taken in a given increment so as not to exceed a predetermined level of sampling uncertainty. One approach is through use of Ingamells's sampling constant (6). Based on the knowledge that the between-sample standard deviation s_s (Equation 1), decreases as the sample size is increased, Ingamells has shown that the relation

$$WR^2 = K_s \qquad (2)$$

is valid in many situations. In Equation 2, W represents the weight of sample analyzed, R is the relative standard deviation (in percent) of sample composition, and K_s is the sampling constant, corresponding to the weight of sample required to limit the sampling uncertainty to 1% with 68% confidence. The magnitude of K_s may be determined by estimating s_s from a series of measurements of samples of weight W.

Once K_s is evaluated for a given sample, the minimum weight W required for a maximum relative standard deviation of R percent can be readily calculated.

An example of an Ingamells sampling constant diagram is shown in Figure 3 for a human liver sample under study in the National Environmental Specimen Bank Pilot Program at the National Bureau of Standards (NBS) in conjunction with the Environmental Protection Agency (7). A major goal of the program is to evaluate specimen storage under different conditions. This requires analysis of small test portions of individual liver specimens. The material must be sufficiently homogeneous that variability between test portions does not mask small variations in composition owing to changes during storage. The homogeneity of a liver sample for sodium was assessed by a radiotracer study in which a portion was irradiated, added to the remainder of the specimen, and the material homogenized. Several test portions were then taken and the activity of ^{24}Na measured as an indicator of the distribution of sodium in the samples. From Figure 3 it can be seen that the weight of sample required to yield an inhomogeneity of 1% (± 2.4 counts g^{-1}s^{-1}) is about 35 g. For a subsample of one gram, a sampling uncertainty of about 5% can be expected.

Minimum Number of Individual Increments. Unless the population is known to be homogeneous, or unless a representative sample is mandated by some analytical problem, sufficient replicate samples (increments) must be analyzed. To determine the minimum number of sample increments, a sampling variance is first obtained, either from previous information on the bulk material or from measurements made on the samples. The number of samples necessary to achieve a given level of confidence can be estimated from the relation

$$n = \frac{t^2 s_s^2}{R^2 \bar{x}^2} \qquad (3)$$

where t is the student's t-table value for the level of confidence desired, s_s^2 and $\bar{x}$ are estimated from preliminary measurements on or from previous knowledge of the bulk material, and R is the percent relative standard deviation acceptable in the average. Initially t can be set at 1.96 for 95% confidence limits and a preliminary value of n calculated. The t value for this n can then be substituted and the system iterated to constant n. This expression is applicable if the sought-for component is distributed in a positive binomial, or a Gaussian, distribution. Such distributions are characterized by having an average, μ, that is larger than the variance, σ_s^2. Remember that values of σ_s (and s_s) may depend greatly on the size of the individual samples.

Two other distributions that may be encountered, particularly in biological materials, should be mentioned. One is the Poisson distribution, in which the sought-for substance is distributed randomly in the bulk material such that σ_s^2 is approximately equal to μ. In this case

$$n = \frac{t^2}{R^2 \bar{x}} \qquad (4)$$

The other is the negative binomial distribution, in which the sought-for substance occurs in clumps or patches, and σ_s^2 is larger than μ. This pattern often occurs in the spread of contamination or contagion from single sources, and is characterized by two factors, the average, $\bar{x}$, and a term, k, called the index of clumping. For this system

$$n = \frac{t^2}{R^2}\left[\frac{1}{\bar{x}} + \frac{1}{k}\right] \quad (5)$$

Here k must be estimated, along with $\bar{x}$, from preliminary measurements on the system.

Sometimes, what is wanted is not an estimate of the mean but instead the two outer values or limits that contain nearly all of the population values. If we know the mean and standard deviation, then the intervals $\mu \pm 2\sigma$ and $\mu \pm 3\sigma$ contain 95% and 99.7%, respectively, of all samples in the population. Ordinarily, the standard deviation σ is not known but only its estimate s, based on n observations. In this case we may calculate statistical tolerance limits of the form $\bar{x} + Ks$ and $\bar{x} - Ks$, with the factor K chosen so that we may expect the limits to include at least a fraction P of the samples with a stated degree of confidence. Values for the factor K (8) depend upon the probability γ of including the proportion P of the population, and the sample size, n. Some values of K are given in Table 1. For example, when $\gamma = 0.95$ and $P = 0.95$, then $K = 3.38$ when $n = 10$, and $K = 37.67$ for duplicates ($n = 2$).

Sampling a Segregated (Stratified) Material. Special care must be taken when assessing the average amount of a substance distributed throughout a bulk material in a nonrandom way. Such materials are said to be segregated. Segregation may be found, for example, in ore bodies, in different production batches in a plant, or in samples where settling is caused by differences in particle size or density.

The procedure for obtaining a valid sample of a stratified material is as follows (9):
- Based on the known or suspected pattern of segregation, divide the material to be sampled into real or imaginary segments (strata).
- Further divide the major strata into real or imaginary subsections and select the required number of samples by chance (preferably with the aid of a table of random numbers).
- If the major strata are not equal in size, the number of samples taken from each stratum should be proportional to the size of the stratum.

In general, it is better to use stratified random sampling rather than unrestricted random sampling, provided the number of strata selected is not so large that only one or two samples can be analyzed from each stratum. By keeping the number of strata sufficiently small that several samples can be taken from each, possible variations within the parent population can be detected and assessed without increasing the standard deviation of the sampling step.

Minimum Number of Individual Increments. When a bulk material is highly segregated, a large number of samples must be taken from different segments. A useful guide to estimating the number of samples to be collected is given by Visman (10), who proposed that the variance in sample composition depends on the degree of homogeneity within a given sample increment and the degree of segregation between sample increments according to the relation

$$s_s^2 = A/W + B/n \quad (6)$$

where s_s^2 is the variance of the average of n samples using a total weight W of sample, and A and B are constants for a given bulk material. A is called a homogeneity constant, and can be calculated from Ingamells's sampling constant and the average composition by

$$A = 10^4 \bar{x}^2 K_s \quad (7)$$

Sampling Materials in Discrete Units. If the lot of material under study occurs in discrete units, such as truckloads, drums, bottles, tank cars, or the like, the variance of the analytical result is the sum of three contributions: (1) that from the variance between units in the lot, (2) that from the average variance of sets of samples taken from within one unit, and (3) that from the variance of the analytical operations. The contribution from each depends upon the number of units in the lot and the number of samples taken according to the following relation (9):

$$\sigma_s^2 = \frac{\sigma_b^2(N - n_b)}{n_b N} + \frac{\sigma_w^2}{n_b n_w} + \frac{\sigma_t^2}{n_t} \quad (8)$$

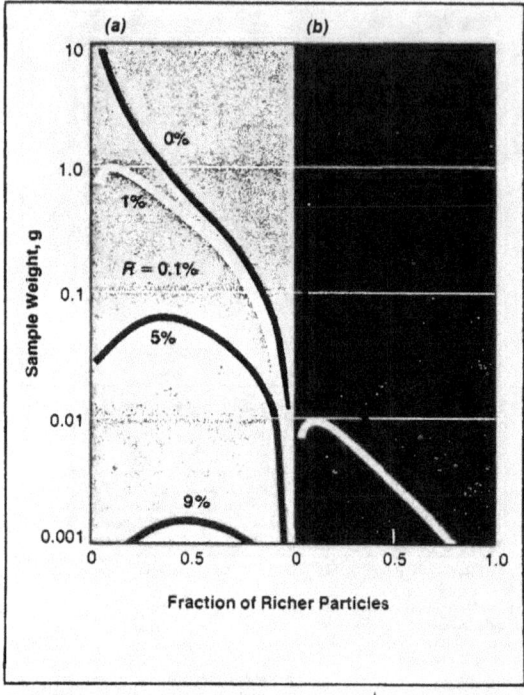

Figure 4. Relation between minimum sample size and fraction of the richer particles in a mixture of two types of spherical particles (diameter 0.1 mm and density 1) for a sampling standard deviation (R) of (a) 0.1% and (b) 1%. Richer particles contain 10% of substance of interest, and leaner ones contain 0, 1, 5, or 9% (after Reference 12, p 554)

Glossary

Bulk sampling—sampling of a material that does not consist of discrete, identifiable, constant units, but rather of arbitrary, irregular units.

Composite—a sample composed of two or more increments.

Gross sample (also called bulk sample, lot sample)—one or more increments of material taken from a larger quantity (lot) of material for assay or record purposes.

Homogeneity—the degree to which a property or substance is randomly distributed throughout a material. Homogeneity depends on the size of the units under consideration. Thus a mixture of two minerals may be inhomogeneous at the molecular or atomic level, but homogeneous at the particulate level.

Increment—an individual portion of material collected by a single operation of a sampling device, from parts of a lot separated in time or space. Increments may be either tested individually or combined (composited) and tested as a unit.

Individuals—conceivable constituent parts of the population.

Laboratory sample—a sample, intended for testing or analysis, prepared from a gross sample or otherwise obtained. The laboratory sample must retain the composition of the gross sample. Often reduction in particle size is necessary in the course of reducing the quantity.

Lot—a quantity of bulk material of similar composition whose properties are under study.

Population—a generic term denoting any finite or infinite collection of individual things, objects, or events in the broadest concept; an aggregate determined by some property that distinguishes things that do and do not belong.

Reduction—the process of preparing one or more subsamples from a sample.

Sample—a portion of a population or lot. It may consist of an individual or groups of individuals.

Segment—a specifically demarked portion of a lot, either actual or hypothetical.

Strata—segments of a lot that may vary with respect to the property under study.

Subsample—a portion taken from a sample. A laboratory sample may be a subsample of a gross sample; similarly, a test portion may be a subsample of a laboratory sample.

Test portion (also called specimen, test specimen, test unit, aliquot)—That quantity of a material of proper size for measurement of the property of interest. Test portions may be taken from the gross sample directly, but often preliminary operations, such as mixing or further reduction in particle size, are necessary.

where

σ_x^2 = variance of the mean,
σ_b^2 = variance of the units in the lot,
σ_s^2 = average variance of the samples taken from a segment,
σ_l^2 = variance of the analytical operations,
N = number of units in the lot,
n_b = number of randomly selected units sampled,
n_a = number of randomly drawn samples from each unit selected for sampling, and
n_t = total number of analyses, including replicates, run on all samples.

If stratification is known to be absent, then much measurement time and effort can be saved by combining all the samples and mixing thoroughly to produce a composite sample for analysis. Equation 8 is applicable to this situation also. If the units vary significantly in weight or volume, the results for those units should be weighted accordingly.

For homogeneous materials σ_a^2 is zero, and the second term on the right-hand side of Equation 8 drops out. This is the case with many liquids or gases. Also, if all units are sampled, then $n_b = N$ and the first term on the right-hand side of Equation 8 also drops out.

Particle Size in Sampling Particulate Mixtures

Random sampling error may occur even in well-mixed particulate mixtures if the particles differ appreciably in composition and the test portion contains too few of them. The problem is particularly important in trace analysis, where sampling standard deviations may quickly become unacceptably large. The sampling constant diagram of Ingamells and the Visman expression are useful aids for estimating sample size when preliminary information is available. Another approach that can often provide insight is to consider the bulk material as a two-component particulate mixture, with each component containing a different percentage of the analyte of interest (11). To determine the weight of sample required to hold the sampling standard deviation to a preselected level, the first step is to determine the number of particles n. The value of n may be calculated from the relation

$$n = \left[\frac{d_1 d_2}{\bar{d}^2}\right]^2 \left[\frac{100(P_1 - P_2)}{R\bar{P}}\right]^2 p(1-p) \tag{9}$$

where d_1 and d_2 are the densities of the two kinds of particles, $\bar{d}$ is the density of the sample, P_1 and P_2 are the percentage compositions of the component of interest in the two kinds of particles, $\bar{P}$ is the overall average composition in percent of the component of interest in the sample, R is the percent relative standard deviation (sampling error) of the sampling operation, and p and $1 - p$ are the fractions of the two kinds of particles in the bulk material. With knowledge of the density, particle diameter, and n, the weight of sample required for a given level of sampling uncertainty can be obtained through the expression, weight = $(4/3)\pi r^3 dn$ (assuming spherical particles).

Figure 4 shows the relation between the minimum weight of sample that should be taken and the composition of mixtures containing two kinds of particles, one containing 10% of the sought-for substance and the other 9, 5, 1, or 0%. A density of 1, applicable in the case of many biological materials, is used, along with a particle diameter of 0.1 mm. If half the particles in a mixture contain 10% and the other half 9% of the substance of interest, then a sample of 0.0015 g is required if the sampling standard deviation is to be held to a part per thousand. If the second half contains 5%, a sample of 0.06 g is necessary; if 1%, 0.35 g would be needed. In such mixtures it is the relative difference in composition that is important. The same sample weights would be required if the compositions were 100% and 90, 50, 10%, or if they were 0.1% and 0.09, 0.05, or 0.01%. The same curves can be used for any relative composition by substitution of x for 10%, and 0.1 x,

B-7

0.5 x, and 0.9 x for the curves corresponding to 1, 5, and 9% in Figure 4. If a standard deviation of 1% is acceptable, the samples can be 100 times smaller than for 0.1%.

An important point illustrated by the figure is that if the fraction of richer particles is small, and the leaner ones contain little or none of the substance of interest, large test portions are required. If a sample of gold ore containing 0.01% gold when ground to 140 mesh (0.1 mm in diameter) consists, say, of only particles of gangue and of pure gold, test portions of 30 g would be required to hold the sampling standard deviation to 1%. (An ore density of 3 is assumed.)

Concluding Comments

Sampling is not simple. It is most important in the worst situations. If the quantities $\bar{x}$, s, K_s, A, and B are known exactly, then calculation of the statistical sampling uncertainty is easy, and the number and size of the samples that should be collected to provide a given precision can be readily determined. But if, as is more usual, these quantities are known only approximately, or perhaps not at all, then preliminary samples and measurements must be taken and on the basis of the results more precise sampling procedures developed. These procedures will ultimately yield a sampling plan that optimizes the quality of the results while holding down time and costs.

Sampling theory cannot replace experience and common sense. Used in concert with these qualities, however, it can yield the most information about the population being sampled with the least cost and effort. All analytical chemists should know enough sampling theory to be able to ask intelligent questions about the samples provided, to take subsamples without introducing additional uncertainty in the results and, if necessary, to plan and perform uncomplicated sampling operations. It is the capability of understanding and executing all phases of analysis that ultimately characterizes the true analytical chemist, even though he or she may possess special expertise in a particular separation or measurement technique.

References

(1) W. J. Youden, *J. Assoc. Off. Anal. Chem.*, 50, 1007 (1967).
(2) T. B. Whitaker, J. W. Dickens, and R. J. Monroe, *J. Am. Oil Chem. Soc.*, 51, 214 (1974); T. B. Whitaker, *Pure Appl. Chem.*, 49, 1709 (1977).
(3) Hazardous Waste Monitoring System, General, *Fed. Regist.*, Vol. 45, No. 98, pp 33075–33127 (May 19, 1980).
(4) Staff, ACS Subcommittee on Environmental Analytical Chemistry, *Anal. Chem.*, 52, 2242 (1980).
(5) See, for example, W. J. Dixon and F. J. Massey, Jr., "Introduction to Statistical Analysis," 3rd ed., McGraw-Hill, New York, 1969; M. G. Natrella, "Experimental Statistics," National Bureau of Standards Handbook 91, August 1963, U.S. Government Printing Office.
(6) C. O. Ingamells and P. Switzer, *Talanta*, 20, 547 (1973); C. O. Ingamells, *Talanta*, 21, 141 (1974); 23, 263 (1976).
(7) S. H. Harrison and R. Zeisler, NBS Internal Report 80-2164, C. W. Reimann, R. A. Velapoldi, L. B. Hagan, and J. K. Taylor, Eds., U.S. National Bureau of Standards, Washington, D.C., 1980, p 66.
(8) M. G. Natrella, "Experimental Statistics," National Bureau of Standards Handbook 91, August 1963, U.S. Government Printing Office, pp 2-13 and Table A6.
(9) ASTM E-300 Standard Recommended Practice for Sampling Industrial Chemicals, American Society for Testing and Materials, Philadelphia, 1973 (reapproved 1979).
(10) J. Visman, *Materials Research and Standards*, November, p 8 (1969).
(11) A. Benedetti-Pichler, in "Physical Methods of Chemical Analysis," W. M. Berl, Ed., Academic Press, New York, 1956, Vol. 3, p 183; W. E. Harris and B. Kratochvil, *Anal. Chem.*, 46, 313 (1974).
(12) W. E. Harris and B. Kratochvil, "Introduction to Chemical Analysis," Saunders, Philadelphia, 1981, Chapter 21.

Kratochvil   *Taylor*

Byron Kratochvil, professor of chemistry at the University of Alberta, received his BS, MS, and PhD degrees from Iowa State University. His research interests include solvent effects on solute properties and reactions, applications of nonaqueous systems to chemical analysis, and methods for determining ionic solutes.

John K. Taylor, coordinator for quality assurance and voluntary standardization activities at the National Bureau of Standards Center for Analytical Chemistry, received his BS from George Washington University, and his MS and PhD degrees from the University of Maryland. His research interests include electrochemical analysis, refractometry, isotope separations, standard reference materials, and the application of physical methods to chemical analysis.

APPENDIX C

Guidelines for Evaluating
The Blank Correction

John Mandel
National Bureau of Standards

The reagent blank, including control of its value by suitable treatment may significantly affect what the measurement is, particularly at the lowest concentration that can be reliably measured. Therefore, a careful control of the analysis of blank values is the order of the day. The present paper discusses the statistical considerations in carrying out blank corrections.

Ordinarily, a measurement involves several consecutive measurements and the measured value of the corresponding background signal, the so-called chemical blank. This is subtracted from the measured value to obtain the "true" concentration of the sample. It is understood that the blank measurement must be properly made and that the resulting correction be correctly applied.

The following should clarify the concept of the blank correction and the proper use of a reagent blank.

Let $\bar{C}_m$ = mean of m measurements of the combined blank of the measurement of the sample, with standard deviation s_m.

$\bar{C}_b$ = mean of b measurements of the concentration of the reagent in the blank, with standard deviation s_b.

$\bar{C}_s$ = best estimate of the concentration of the measurement in the sample, corrected for the blank.

The statistical uncertainty of $\bar{C}_m$ is given by

$$\pm \frac{t s_m}{\sqrt{m}}$$

where $t = t_{1-\alpha/2}$ is the t value for $m-1$ degrees of freedom for the $100(1-\alpha)$ confidence level.

Likewise the statistical uncertainty of $\bar{C}_b$ is given by

$$\pm \frac{t s_b}{\sqrt{b}}$$

The uncertainty of $\bar{C}_s$ is obtained by combination of the two.

$$\bar{C}_s \pm t\sigma_s = (\bar{C}_m - \bar{C}_b) \pm t \sqrt{\frac{s_m^2}{m} + \frac{s_b^2}{b}} \qquad (1)$$

where $t = t_{1-\alpha/2}$ is the t value for f* degrees of freedom (see equation 3) at the $100(1-\alpha)\%$ confidence level.

In the case where the measurement system is demonstrated to be in a state of statistical control and the respective standard deviations are known, equation (1) becomes

$$\bar{C}_s \pm Z\sigma_s = (\bar{C}_m - \bar{C}_b) \pm Z \sqrt{\frac{\sigma_m^2}{m} + \frac{\sigma_b^2}{b}} \qquad (1a)$$

Where $Z = Z_{1-\alpha/2}$ ≈ 1.96 for the 95 percent confidence interval ($\alpha = 0.05$).

A special case exists when $\bar{C}_m \approx \bar{C}_b$. In this case $s_b \approx s_m = s$ so that

$$\bar{C}_s = \bar{C}_m - \bar{C}_b \pm ts \sqrt{\frac{m+b}{mb}} \qquad (2)$$

where $t = t_{1-\alpha/2}$ is the t value for m+b-2 degrees of freedom for the $100(1-\alpha)\%$ confidence level and

$$s = \sqrt{\frac{(m-1)s_m^2 + (b-1)s_b^2}{m+b-2}}$$

in the case of statistical quality control with $\sigma_m = \sigma_b = \sigma$, one may use

$$\bar{C}_s = (\bar{C}_m - \bar{C}_b) \pm Z_{1-\alpha/2} \, \sigma \sqrt{\frac{m+b}{mb}} \qquad (2a)$$

The expressions (2) and (2a) are based on measurement of the blank and sample by the same method and apply even if the measurand is not detected in the blank. If the blank and sample are measured by different methods, then the equations (1) or (1a) apply.

Appropriate values of t, based upon the effective degrees of freedom, f*, must be used in equation (1). These may be computed from the equation [2].

$$f^* = \left(\frac{(V_m + V_b)^2}{\frac{V_m^2}{m+1} + \frac{V_b^2}{b+1}} \right) - 2 \qquad (3)$$

In equation (3), the variance, V, signifies s^2.

Whenever the blank correction becomes significant, it is necessary to measure it with sufficient care. It is clear from the above that blank measurements may need to be made with the same amount of effort as the sample, itself, as $C_m \approx C_b$. This fact is often overlooked by experimenters who may make a limited number of measurements of the blank while devoting most of their effort to measurement of the sample.

Blank corrections become increasingly important in the case of measurements close to the limit of detection. The effect of small variability of the blank is magnified in this case. Likewise even small constant blanks can result in the differencing of two quantities approaching each other in magnitude.

The question of acceptable limits for the blank will now be addressed. The absolute value of the blank would appear to be less important than its accurate evaluation. However, it is a necessary correction and good measurement practice dictates that it should be kept within reasonable limits. An empirical rule in the case of trace analysis is to limit the blank correction to no more than ten times the acceptable limit of error for the measurement and furthermore that it should never exceed the concentration level expected in the sample. The logic behind the first condition is that up to a ten percent error in estimation of the blank would cause no serious difficulties. The second condition is to prevent minor errors in the two measured quantities from introducing large errors in the difference which is the quantity of practical interest.

In the preceding discussion, the significance of the blank was considered on the basis of its contribution as a concentration factor under the final conditions of measurement. Furthermore, the term C_b contains the sum of the contributions from each source of blank. If C_b is excessive, and if several sources (reagents) are involved, measurements must be carried out in a suitable program to identify each source and the magnitude of its contribution in order to take corrective actions. Obviously, the magnitude of each source of blank and the final conditions of measurement (e.g., final volume of a solution) must be considered in establishing a permissible level for the blank for each reagent used.

In all of the above discussions, it was assumed that the blank measurement simulates the sample measurement process so that the value of C_b is meaningful. In some cases it may be difficult or even impossible to fully simulate the sample measurement process unless the sample matrix is present in critical steps of the procedure. If matrix simulation is necessary and cannot be achieved, it may be necessary to independently analyze each reagent for its measurand content and calculate its contribution to the measurement blank.

A related question is the uncertainty of the measured values resulting from uncertainties in the analytical function. Most measurements involve the use of such a function to relate the measured quantity (signal) to the concentration of the sample. Uncertainties in this function must be considered as a measurement uncertainty. The uncertainty in the analytical function is not a significant consideration in the blank correction, provided both measurements use the same function. However, it must be considered in evaluating the final measured value.

References

[1] Thomas J. Murphy, "The Role of the Analytical Blank in Accurate Trace Analysis", NBS Special Publication 422, Accuracy in Trace Analysis: Sampling, Sample Handling, Analysis (1976).

[2] Mary G. Natrella, "Experimental Statistics", NBS Handbook 91, p. 3-28, (1963).

APPENDIX D

Report

John K. Taylor
Center for Analytical Chemistry
National Bureau of Standards
Washington, D.C. 20234

Validation of Analytical Methods

Validation of analytical methods is a subject of considerable interest. Documents such as the "ACS Guidelines for Data Acquisition and Data Quality Evaluation" (1) recommend the use of validated methods. The promulgation of federal environmental regulations requires the inclusion of validated reference methods. Standards-writing organizations spend considerable time in collaborative testing of methods they prepare, validating them in typical applications and determining their performance characteristics. Nevertheless, questions about the appropriateness of methods and the validity of their use in specific situations often arise. Some of these questions may be due to differences in understanding both what a method really is and what the significance of the validation process is. This paper attempts to clarify the nomenclature of analytical methodology and to define the process of validating methods for use in specific situations.

Hierarchy of Methodology

The hierarchy of methodology, proceeding from the general to the specific, may be considered as follows:
technique → method → procedure → protocol.

A *technique* is a scientific principle that has been found to be useful for providing compositional information; spectrophotometry is an example. Analytical chemists historically have investigated new measurement techniques for their ability to provide novel measurement capability, or to replace or supplement existing methodology. As a result of innovative applications, analysts can now analyze

for myriad substances in exceedingly complex mixtures at ever lower trace levels, with precision and accuracy undreamed of only a few years ago (2).

A *method* is a distinct adaptation of a technique for a selected measurement purpose. The pararosaniline method for measurement of sulfur dioxide is an example. It involves measuring the intensity of a specific dye, the color of which is "bleached" by the gas. Several procedures for carrying out this method may be found in the literature. Modern methodology is complex and may incorporate several measurement techniques and thus be interdisciplinary.

A *procedure* consists of the written directions necessary to utilize a method. The "standard methods" developed by ASTM and AOAC are, in reality, standardized procedures. ASTM D2914—Standard Test Method for the Sulfur Dioxide Content of the Atmosphere (West-Gaeke Method) is an example (3). While a precise description is the aim, it is difficult, if not impossible, to describe every de-

Hierarchy of Analytical Methodology

	Definition	Example
Technique	Scientific principle useful for providing compositional information	Spectrophotometry
Method	Distinct adaptation of a technique for a selected measurement purpose	Pararosaniline method for measurement of sulfur dioxide
Procedure	Written directions necessary to use a method	ASTM D2914—Standard Test Method for the Sulfur Dioxide Content of the Atmosphere (West-Gaeke Method)
Protocol	Set of definitive directions that must be followed, without exception, if the analytical results are to be accepted for a given purpose	EPA Reference Method for the Determination of Sulfur Dioxide in the Atmosphere (Pararosaniline Method)

The Report is based on a talk given at the 184th ACS National Meeting, Sept. 12–17, 1982, Kansas City, Mo.

tail of every operation in a procedure. Accordingly, some level of sophistication is presumed for the user of every published procedure, if very sophisticated users are contemplated, only a minimum of detail will be provided and vice versa. However, it should be noted that any omission in the description of critical steps is a potential source of variance or bias, even in the hands of knowledgeable analysts. Because of the flexibility intentionally or unintentionally provided to the analyst, or because of differences in interpretation, it is fair to say that numerous differences of application occur in the use of even the most precisely defined procedures. Such differences often account for the interlaboratory variability observed in many collaborative tests. Further, at some point of departure from a published procedure, a new method results that may need its own validation.

The term *protocol* is the most specific name for a method. A protocol is a set of definitive directions that must be followed, without exception, if the analytical results are to be accepted for a given purpose. Protocols may consist of existing methods or procedures, modifications of such, or they may be developed especially for specific purposes. Typically, they are prescribed by an official body for use in a given situation such as a regulatory process. The EPA Reference Method for the Determination of Sulfur Dioxide in the Atmosphere (Pararosaniline Method) is an example of a protocol (1). The test method specified as part of a contractual arrangement for the acceptance of data or a product or material is another example of a protocol, although it may not be called that in the contract.

A plethora of methods, procedures,

and protocols based on the same measurement principle can arise for a given analytical determination. Usually, they are worded differently, and they may contain subtle or major differences in treatment of details. The extent to which the documents to be individually considered as matters of professional judgement. It is evident that some variations could be merely a method of experimentally testing the clarity of the written word.

Goals for Validation

Validation is the process of determining the suitability of methodology as providing useful analytical data. This is the judgement in which the performance parameters of the method of concern relate to the requirements for the analytical data, as illus-

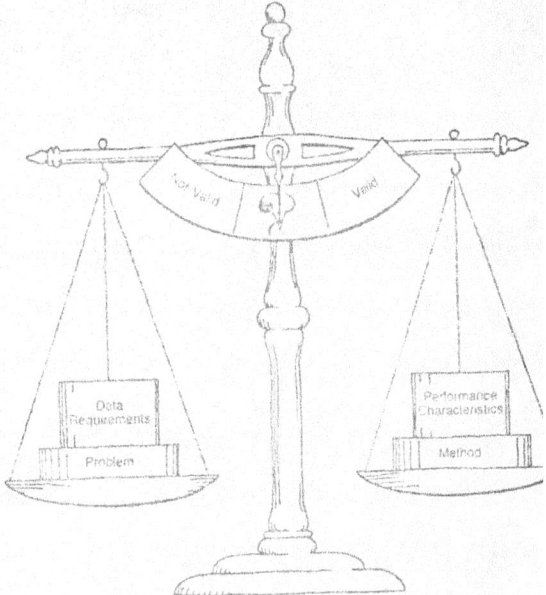

Figure 1. Basic concept of the validation process

trated in Figure 1. Obviously, a method that is valid in one situation could be invalid in another. Accordingly, the establishment of firm requirements for the data is a prerequisite for method selection and validation. When data requirements are ill considered, analytical measurement can be unnecessarily extensive or the method chosen is more accurate than required, inadequate, or the method is less accurate than required, or futile if the accuracy of the method is unknown.

Fortunately, typical analytical measurement problems often encountered include a wide variety of field analyses, environmental determinations, and routine measurement for the characterization of industrial product. The kinds of samples for which methods have been validated.

tail of every operation in a procedure. Accordingly, some level of sophistication is presumed for the user of every published procedure; if very sophisticated users are contemplated, only a minimum of detail will be provided and vice versa. However, it should be noted that any omission in the description of critical steps is a potential source of variance or bias, even in the hands of knowledgeable analysts. Because of the flexibility intentionally or unintentionally provided to the analyst, or because of differences in interpretation, it is fair to say that minor or major differences of application occur in the use of even the most precisely defined procedures. Such differences often account for the interlaboratory variability observed in many collaborative tests. Further, at some point of departure from a published procedure, a new method results that may need its own validation.

The term *protocol* is the most specific name for a method. A protocol is a set of definitive directions that must be followed, without exception, if the analytical results are to be accepted for a given purpose. Protocols may consist of existing methods or procedures, modifications of such, or they may be developed especially for a specific purpose. Typically, they are prescribed by an official body for use in a given situation such as a regulatory process. The EPA Reference Method for the Determination of Sulfur Dioxide in the Atmosphere (Pararosaniline Method) is an example of a protocol (4). The test method specified as part of a contractual arrangement for the acceptance of data or a product or material is another example of a protocol, although it may not be called that in the contract.

A plethora of methods, procedures, and protocols based on the same basic statement principle can arise for a given analytical determination. Used as they are worded differently, they may give contradictable or minor differences in technical detail. The extent to which one needs to be distinguished from another is a matter of professional judgment. It is evident that some analytical tests could be merely a matter of experimentally testing the fidelity of the written word.

Goals for Validation

Validation is the process of determining the suitability of methodology for providing useful analytical data. This implies establishment of which the performance parameters of the method and the limitations which are requirements for the analytical data to be illustrated in Figure 1. Obviously, a method that is valid in one situation could be invalid in another. Accordingly, the establishment of data requirements is a prerequisite for method selection and validation. When data requirements are ill considered, analytical measurements can be unnecessarily expensive or the method chosen is more elaborate than required, inadequate if the method is less accurate than required, or utterly futile if the accuracy of the method is unknown.

Unfortunately, analytical and even standards development scientists often neglect to consider this in a wide variety of physical and chemical environmental determinations and/or the measurement of one characterization of industrial products. The kinds of samples for which methods have been val-

Figure 1. Basic concept of the validation process

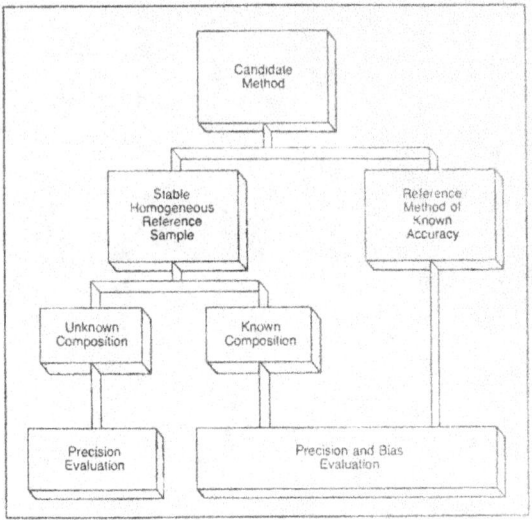

Figure 2. Collaborative test process

ultimate use of analytical methodology is to produce compositional information about specific samples necessary for the solution of particular problems ranging from exotic research investigations to the very mundane. The selection of appropriate measurement methodology is often a major consideration. Methods or procedures, even if previously validated in general terms, cannot unequivocally be assumed to be valid for the situation in hand, because of possible differences in sample matrix and other considerations. Professional analytical chemists traditionally have recognized this and their responsibility to confirm or prove (if necessary) both the validity of the methodology used for specific application (2) and their own ability to reduce it to practice.

The classical validation process is illustrated in Figure 2. When reference samples are available that are similar in all respects to the test samples, the process is very simple. It consists of analyzing a sufficient number of reference samples and comparing the results to the expected or certified values (7). Before or during such an exercise, the analyst must demonstrate the attainment of a state of statistical control of the measurement system (8) so that the results can be relied upon as representative of those expected when using the methodology-measurement system.

When a suitable reference material is not available, several other approaches are possible. One consists of comparing the results of the candidate method with those of another method known to be applicable and reliable, but not useful in the present situation because of cost, unavailability of personnel or equipment, or other reasons. Even the agreement of results with those obtained using any additional independent method can provide some useful information.

Spiked samples and surrogates may be used as reference samples. This approach is less desirable and less satisfactory because of the difficulty in the reliable preparation of such samples and because artificially added materials such as spikes and surrogates may exhibit matrix effects differing from those of natural samples. Split samples of the actual test samples may be used to evaluate the precision of a method or procedure, but they provide no information about the presence or magnitude of any measurement bias.

Another approach is to infer the appropriateness of methodology from measurements on analogous but dissimilar reference materials. The critical professional judgment of the analyst is necessary to decide the validity of the inference.

In all cases, sufficient tests must be made to evaluate the methodology for

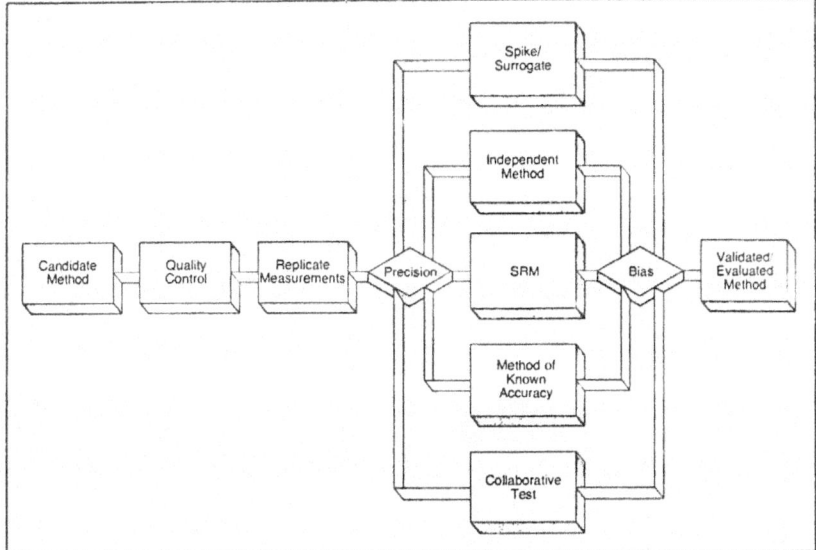

Figure 3. General process for evaluation/validation of methodology

the variety of matrices and ranges of composition expected during the measurement process. Ordinarily, the latter should include three levels of concentration, namely, the extremes and the mid-range of compositions expected. Statistical considerations suggest that at least six degrees of freedom (ordinarily seven measurements) should be involved at each decision point.

Conclusion

A valid method is necessary but not sufficient for the production of valid data. Most methods require a degree of skill on the part of the analyst; this skill constitutes a critical factor in the measurement process. It is common knowledge that data obtained by several laboratories on the same test sample using the same methodology may show a high degree of variability. The alleviation of such a problem is in the area of quality assurance of the measurements (8). Data obtained by a valid method used in a well-designed quality assurance program should allow the assignment of limits of uncertainty that can be used to judge the data's validity.

It should be remembered that the validity of any data will also depend upon the validity of the model and of the sample (8, 9). The model represents the conceptualization of the problem to be solved, describes the samples that should be analyzed, the data base required, and the way the model will be utilized. Obviously, even flawless measurement data will be of little value if the basic concepts are faulty. Likewise the samples analyzed must be valid if the results obtained for them are to be intelligently interpreted.

The key role of reliable reference materials in the validation of analytical measurements cannot be overemphasized. Their use in validating the methodology has already been discussed. A planned sequential analysis of reference materials in a quality assurance program can assess the quality of the data output and thus validate the overall aspects of the analytical measurement system (7).

References

(1) ACS Subcommittee on Environmental Analytical Chemistry. *Anal. Chem.* 1980, 52, 2242.
(2) Taylor, John K. CHEMTECH 1982, 12, 225.
(3) Annual Book of ASTM Standards, Part 26. American Society for Testing and Materials, Philadelphia, Pa. 19103, 1981.
(4) National Primary and Secondary Ambient Air Quality Standards, *Fed. Regist.* April 30, 1971, Vol. 36, Part 84, p. 8187.
(5) Glaser, J. A.; Foerst, D. L.; McKee, G. D.; Quave, S. A.; Budde, W. L. *Environ. Sci. Technol.* 1981, 15, 1426–35.
(6) Purdue, L. T. et al. *Environ. Sci. Tech-*
nol. 1972, 6, 152–54.
(7) Taylor, John K. "Reference Materials—How They Are or How They Should Be Used," *ASTM Journal of Testing and Technology*, in press.
(8) Taylor, John K. *Anal. Chem.* 1981, 53, 1588–96 A.
(9) Kratochvil, Byron; Taylor, John K. *Anal. Chem.* 1981, 53, 924–38 A.

John K. Taylor is coordinator for chemical measurement assurance and voluntary standardization at the National Bureau of Standards' Center for Analytical Chemistry. He received his BS degree from George Washington University and his MS and PhD degrees from the University of Maryland. His research interests include electrochemical analysis, refractometry, isotope separations, standard reference materials, and the application of physical methods to chemical analysis.

APPENDIX E

E. Selected References

1. ASTM D2777, ASTM Philadelphia, Pa 19103. Standard Practice for determination of the precision and bias of methods of committee D19 on water.
2. ASTM D3614, ASTM Philadelphia, Pa 19103. Evaluating laboratories engaged in sampling and analysis of atmospheres and emissions.
3. ASTM D3856, ASTM Philadelphia, Pa 19103. Evaluating laboratories engaged in sampling and analysis of water and waste-waters.
4. ASTM E178, ASTM Philadelphia, Pa 19103. Standard recommended practice for dealing with outlying observations.
5. ASTM E305-83, ASTM Philadelphia, PA 19103. Standard practice for establishing and controlling spectrochemical analytical curves.
6. ASTM E-548, ASTM Philadelphia, PA 19103. Standard recommended practice for generic criteria for use in evaluation of testing and/or inspection agencies.
7. ASTM E-748, ASTM Philadelphia, PA 19103. Quality assurance procedures for spectrographic laboratories.
8. G. A. Berman, Ed., Testing laboratory performance: Evaluation and accreditation. NBS special publication 591.
9. R. Bordner and J. Winter, Microbiological methods for monitoring the environment, water, and wastewater. EPA-600/8-78-017, Dec. 1978.
10. J. Bryson, et al, Bibliography on laboratory accreditation, NBSIR 82-2523 (1982) National Bureau of Standards, Washington, D.C. 20234.
11. J. P. Cali, et al, The role of standard reference materials in measurement systems, NBS monograph 148.

12. ASTM Manual on presentation of data and control chart analysis. STP 15D, ASTM, Philadelphia, PA 19103.
13. C. Eisenhart, Realistic evaluation of the precision and accuracy of instrument calibration systems, in ref. 26, pp 21-47.
14. J. J. Filliben, Testing basic assumptions in the measurement process, in "Validation of the Measurement Process", J. R. Devoe Ed., ACS Symposium Series No. 63 (1977).
15. L. C. Friedman and D. E. Erdmann, Quality assurance practices for the chemical and biological analyses of water and fluvial sediments-Chapter A6 in "Techniques of Water-Resources Investigations of USGS" (1982).
16. S. Gaft and F. D. Richards, Quality assurance at Ford Motor Company Central Laboratory-A dynamic apporch to laboratory quality, in ASTM STP/ 814.
16A. J. A. Glaser, et al., E. S. & T. 15, 1426-35 (1981).
17. Trace analysis for wastewaters - Method detection limit, J. A. Glaser et al. E. S. & T. 15, 1426-35 (1981).
18. Guidelines for data acquisition and data quality evaluation in environmental chemistry, American Chemical Society, Analytical chemistry 52: 2242-2249; 1980.
19. Handbook for analytical quality control in water and wastewater laboratories EPA 600/4-79-019, March 1979.
20. C. D. Hendrix, What every technologist should know about experimental design. CHEMTECH, March 1979, pp. 167-174.
21. H. S. Hertz and C. N. Chesler, Eds. Trace organic analysis: A new frontier in analytical chemistry. NBS special publication 519.
22. J. Stuart Hunter, Calibration and the straight line: Current statistical practices. AOAC journal 64, No. 3, 574-83 (1982).

23. ISO Guide 25, Guidelines for assessing the technical competence of testing laboratories, American National Standards Institute, 1430 Broadway, New York, N.Y. 10018.

24. B. B. Kratochvil and J. K. Taylor, Sampling for chemical analysis, et. al., Anal. Chem. 53, 924A-38 (1981).

25. B. Kratochvil and J. K. Taylor, A survey of the recent literature on sampling for chemical analysis, NBS Technical Note 1153, January 1982, National Bureau of Standards, Washington, D.C. 20234.

26. H. S. Ku, Editor, Precision Measurement and Calibration: Statistical Concepts and Procedures, NBS Special Publication 300, Vol. 1. Stock No. 003-003-00072-8 Superintendent of Documents, U.S. Govt. Printing Office, Washington, D.C. 20402 ($17.00).

27. P. D. LaFleur, Ed., Accuracy in Trace Analysis: Sampling, Sample Handling, Analysis, NBS Special Publication 422.

28. J. Mandel and T. W. Lashof, Interpretation and Generalization of Youden's Two-Sample Diagram, J. Quality Technology 6 pp. 22-36 (1974).

29. A. G. McNish, Dimensions, Units, and standards, Physics Today, 10, pp. 19-25, April 1957.

30. W. W. Meinke, Ed., Trace Characterization, Chemical and Physical, NBS Monograph 100, National Bureau of Standards, Washington, D.C. 20234.

31. M. G. Natrella, Experimental Statistics, NBS Handbook 91, Stock No. 003-003-00135-0, Superintendent of Documents, U.S. Govt. Printing Office, Washington, D.C. 20402 ($18.00).

32. L. S. Nelson, Use of Range to Estimate Variability, J. Qual. Technology, Vol. 7 No. 1, Jan. 1975.

33. W. Nelson, How to Analyze Data with Simple Plots. ASQC Basic References in Quality Control - Statistical Techniques.

34. Applied Linear Statistical Models, J. Nether and W. Wasserman, p. 140-145 (1974). Published by Richard D. Irwin, Inc., Homewood, Illinois 60430.
35. W. E. Oatess - Establishment of Accreditation Programs for Environmental Labs. Env. Sci. Tech. $\underline{12}$ 1124-27 (1978).
36. Precision Measurement and Calibration, NBS Handbook 77.
37. Principles of Environmental Measurement, American Chemical Society, Anal. Chem. 55 2210-18 (1983).
38. Quality Control System - Requirements for a Testing and Inspection Laboratory, American Council of Independent Laboratories, 1725 K Street N.W., Washington, D.C. 20006.
39. Standard Reference Materials and Meaningful Measurement, NBS Special Publication 408.
40. Standard Reference Materials Catalogue, NBS Special Publication 260. National Bureau of Standards Washington, D.C. 20234.
41. J. K. Taylor, Importance of Inter-calibration in Marine Analysis, Thal. Jugo. $\underline{14}$ 221-29 (1978).
42. J. K. Taylor, Quality Assurance of Chemical Measurements, Anal. Chem. $\underline{53}$ 1588A-96A (1981).
43. J. K. Taylor, Quality Assurance Measures for Environmental Data, in "Lead in the Marine Environment", M. Branica and Z. Konrad Eds., Pergamon Press (1980) pp. 1-7.
44. J. K. Taylor, Validation of Analytical Methods, Anal. Chem. $\underline{55}$, 600A (1983).
45. J. K. Taylor, Reference Materials - What they are and how they should be used. J. Testing and Evaluation $\underline{11}$, 355-7 (1983).
46. G. Wernimount - Ruggedness Evaluation of Test Procedures, ASTM Standardization News - pp. 13-16, March 1977.

47. J. O. Westgard and T. Groth, A multi-rule Shewhart Chart for Quality Control in Clinical Chemistry, Clinical Chemistry $\underline{27}$, 495-501 (1981).
48. William J. Youden, Collection of Papers on Statistical Treatment of Data. Journal of Quality Technology, Vol. 4, No. 1, pp. 1-67, January 1972.
49. Ranking Laboratories by Round Robin Tests, W. J. Youden, Materials Research and Standards, Jaunary 1963.
50. W. J. Youden and E. H. Steiner, Statistical Manual of the Association of Official Analytical Chemists, AOAC, PO Box 540, Washington, DC 20044.
51. ASTM E882 Accountability and Quality Control in the Chemical Analysis Laboratory.
52. ASTM E173 Conducting Interlaboratory Studies of Methods of Chemical Analysis of Metals.
53. ASTM C1009 Establishing a Quality Assurance Program for Analytical Chemistry Laboratories Within the Nuclear Industry.
54. CFR Title 21, Food and Drugs, Chapter 1, F&DA, Part 38, GLP's for Non-Clinical Laboratory Studies.
55. CFR Title 40, Part 792, Toxic Substances Control. Fed. Reg. $\underline{48}$, No. 230, November 28, 1983, pp. 53922.
56. CFR Title 42, Public Health, Chapter 1, Public Health Service, Part 74, Clinical Laboratories.
57. QAMS-005/80, Interim Guidelines and Specifications for Preparing Quality Assurance Project Plans (EPA).
58. L. P. Provost, "Statistical Methods in Environmental Sampling" in Environmental Sampling for Hazardous Wastes, ACS Symposium Series 267 (1984) American Chemical Society, Washington, DC 20036.

QUALITY ASSURANCE CODE OF ETHICS

With full appeciation of my responsibilities as an analytical chemist, I subscribe to all apsects of The Chemist's Creed and to the ethical practice of the profession of analytical chemistry and particularly will strive to:

1. Acquire a full understanding and develop peer technical expertise in every area of analytical chemistry in which my professional services are offered.

2. Comprehend, to the extent possible, all problems for which my analytical services are required; assure myself of the validity of the approach selected; understand any limitations on the measurements and discuss such with clients, as appropriate.

3. Use validated methodology, exclusively.

4. Demonstrate statistical control of the measurement system before definitive measurements are made.

5. Calibrate, to the extent necessary and possible; engage in intercalibration activities as appropriate to minimize chance of internal laboratory bias.

6. Utilize recognized Good Laboratory Practices (GLP's) and Good Measurement Practices (GMP's) throughout all aspects of sampling and measurement processes.

7. Utilize documented procedures and record all significant experimental details in such a way that the measurements could be reproduced by myself or a competent analyst at a later date as necessary.

8. Provide or have available limits of uncertainty of all data reported including that due to sample and to measurement, supported by statistical inference and/or professional judgment, as pertinent; state clearly the basis for any interpretations provided of the measurement data.

9. Confirm the qualitative identification of all parameters measured and provide supporting evidence as necessary.

10. Retain all samples, data and documentary evidence as necessary for a period of time commensurate with its importance.

www.ingramcontent.com/pod-product-compliance
Lightning Source LLC
Chambersburg PA
CBHW051633170526
45167CB00001B/172